4

Springer Series on Chemical Sensors and Biosensors

Methods and Applications

Series Editor: O. S. Wolfbeis

Springer Series on Chemical Sensors and Biosensors

Series Editor: O. S. Wolfbeis

Recently Published and Forthcoming Volumes

Piezoelectric Sensors
Volume Editors: Steinem C., Janshoff A.
Vol. 5, 2006

Surface Plasmon Resonance Based Sensors
Volume Editor: Homola J.
Vol. 4, 2006

Frontiers in Chemical Sensors
Novel Principles and Techniques
Volume Editors: Orellana G., Moreno-Bondi M. C.
Vol. 3, 2005

Ultrathin Electrochemical Chemo- and Biosensors
Technology and Performance
Volume Editor: Mirsky V. M.
Vol. 2, 2004

Optical Sensors
Industrial, Environmental and Diagnostic Applications
Volume Editors:
Narayanaswamy R., Wolfbeis O. S.
Vol. 1, 2003

Surface Plasmon Resonance Based Sensors

Volume Editor: Jiří Homola

With contributions by

J. Dostálek · J. Homola · S. Jiang · J. Ladd
S. Löfås · A. McWhirter · D. G. Myszka · I. Navratilova
M. Piliarik · J. Štěpánek · A. Taylor · H. Vaisocherová

Springer

Chemical sensors and biosensors are becoming more and more indispensable tools in life science, medicine, chemistry and biotechnology. The series covers exciting sensor-related aspects of chemistry, biochemistry, thin film and interface techniques, physics, including opto-electronics, measurement sciences and signal processing. The single volumes of the series focus on selected topics and will be edited by selected volume editors. The Springer Series on Chemical Sensors and Biosensors aims to publish state-of-the-art articles that can serve as invaluable tools for both practitioners and researchers active in this highly interdisciplinary field. The carefully edited collection of papers in each volume will give continuous inspiration for new research and will point to existing new trends and brand new applications.

Library of Congress Control Number: 2006924584

ISSN 1612-7617
ISBN-10 3-540-33918-3 Springer Berlin Heidelberg New York
ISBN-13 978-3-540-33918-2 Springer Berlin Heidelberg New York
DOI 10.1007/b100321

Springer is a part of Springer Science+Business Media

springer.com

Cover design: WMXDesign GmbH, Heidelberg
Typesetting and Production: LE-TEX Jelonek, Schmidt & Vöckler GbR, Leipzig

Printed on acid-free paper 02/3100 YL – 5 4 3 2 1 0

Series Editor

Prof. Dr. Otto S. Wolfbeis

Institute of Analytical Chemistry
University of Regensburg
Chemo- and Biosensors
93040 Regensburg, Germany
otto.wolfbeis@chemie.uni-regensburg.de

Volume Editor

Dr. Jiří Homola

Department of Optical Sensors
Institute of Radio Engineering and Electronics
Academy of Sciences of the Czech Republic
Chaberská 57
18251 Prague 8
Czech Republic
homola@ure.cas.cz

Springer Series on Chemical Sensors and Biosensors Also Available Electronically

For all customers who have a standing order to Springer Series on Chemical Sensors and Biosensors, we offer the electronic version via SpringerLink free of charge. Please contact your librarian who can receive a password or free access to the full articles by registering at:

springerlink.com

If you do not have a subscription, you can still view the tables of contents of the volumes and the abstract of each article by going to the SpringerLink Homepage, clicking on "Browse by Online Libraries", then "Chemical Sciences", and finally choose Springer Series on Chemical Sensors and Biosensors.

You will find information about the

- Editorial Board
- Aims and Scope
- Instructions for Authors
- Sample Contribution

at springer.com using the search function.

Preface

Over the last two decades, surface plasmon resonance (SPR) sensors have attracted a great deal of attention. A myriad of research reports have appeared describing advancements in SPR sensor technology and its applications. SPR sensor technology has been commercialized and SPR biosensors have become a central tool for characterizing and quantifying biomolecular interactions.

This book is intended to provide a comprehensive treatment of the field of SPR sensors. It is hoped that the material is sufficiently detailed to be of real value to both people involved directly with SPR sensors and to people using similar sensing methods.

The book is divided into three parts. Part I introduces readers to the fundamental principles of surface plasmon resonance (bio)sensors and covers the electromagnetic theory of surface plasmons, the theory of SPR sensors and includes an analysis of molecular interactions at sensor surfaces. Part II presents a review of the state-of-the-art in the development of two key elements of SPR biosensors: optical instrumentation and functionalization methods. Part III discusses applications of SPR biosensors. The part begins with a chapter devoted to applications of SPR sensors to research in molecular interactions. The following chapters discuss progress towards developing SPR biosensor-based detection systems suitable for field use and applications of SPR biosensors for the detection of chemical and biological analytes related to environmental monitoring, food safety and security, and medical diagnostics.

I would like to thank all of the contributors from around the world who have contributed material to this book. I am also indebted to Prof. Otto S. Wolfbeis, Editor of the Springer Series on Chemical Sensors and Biosensors and Peter W. Enders, Senior Executive Editor, Springer, for their support of this project. My special thanks go to my wife Hana, without whom this book could not have been written.

Prague, July 2006 Jiří Homola

Contents

Part I
Fundamentals of SPR Sensors

Springer Ser Chem Sens Biosens (2006) 4: 3–44
DOI 10.1007/5346_013

Published online: 5 July 2006

Electromagnetic Theory of Surface Plasmons

Jiří Homola

Institute of Radio Engineering and Electronics, Prague, Czech Republic
homola@ure.cas.cz

Keywords Excitation of surface plasmons · Grating coupler · Guided mode · Optical waveguide · Prism coupler · Surface plasmons

1 Introduction

The first documented observation of surface plasmons dates back to 1902, when Wood illuminated a metallic diffraction grating with polychromatic light and noticed narrow dark bands in the spectrum of the diffracted light, which he referred as to anomalies [1]. Theoretical work by Fano [2] concluded that these anomalies were associated with the excitation of electromagnetic surface waves on the surface of the diffraction grating. In 1958 Thurbadar observed a large drop in reflectivity when illuminating thin metal films on a substrate [3], but did not link this effect to surface plasmons. In 1968 Otto explained Turbadar's results and demonstrated that the drop in the reflectivity in the attenuated total reflection method is due to the excitation of surface plasmons [4]. In the same year, Kretschmann and Raether reported excitation of surface plasmons in another configuration of the attenuated total reflection method [5]. The pioneering work of Otto, Kretschmann, and Raether established a convenient method for the excitation of surface plasmons and their investigation, and introduced surface plasmons into modern optics (see, for example [6], and [7]). In the late 1970s, surface plasmons were first employed for the characterization of thin films [8] and the study of processes at metal boundaries [9].

In this chapter we present an electromagnetic theory of surface plasmons based on theoretical analysis of light propagation in planar metal/dielectric waveguides. The main characteristics of surface plasmons propagating along metal–dielectric and dielectric–metal–dielectric waveguides are introduced and methods for optical excitation of surface plasmons are discussed.

2 Theory of Planar Metal/Dielectric Waveguides

In this section, we present an electromagnetic theory of optical waveguides based on solving Maxwell's equations using the modal method [10–12]. In this approach, the electric and magnetic field vectors $\boldsymbol{E}$ and $\boldsymbol{H}$ are each expressed as a sum of field contributions, one part representing power that is guided along the waveguide, the remaining part representing power that is radiated from the waveguide [10]:

$$\boldsymbol{E}(\boldsymbol{r},t) = \boldsymbol{E}_{\mathrm{G}}(\boldsymbol{r},t) + \boldsymbol{E}_{\mathrm{R}}(\boldsymbol{r},t) , \tag{1}$$

$$\boldsymbol{H}(\boldsymbol{r},t) = \boldsymbol{H}_{\mathrm{G}}(\boldsymbol{r},t) + \boldsymbol{H}_{\mathrm{R}}(\boldsymbol{r},t) , \tag{2}$$

where subscript G and R denote the guided and radiation fields, $\boldsymbol{r}$ is space vector and t is time. The guided, or bound, portion can be expressed as a finite sum of guided modes:

$$\boldsymbol{E}_{\mathrm{G}}(\boldsymbol{r},t) = \sum_j \alpha_j \boldsymbol{E}_j(\boldsymbol{r},t) , \tag{3}$$

$$\boldsymbol{H}_{\mathrm{G}}(\boldsymbol{r},t) = \sum_j \alpha_j \boldsymbol{H}_j(\boldsymbol{r},t) , \tag{4}$$

where j is a mode number ($j = 1, 2, ..., M$) and α_j are modal amplitudes. The modal fields $\boldsymbol{E}_j(\boldsymbol{r},t)$ and $\boldsymbol{H}_j(\boldsymbol{r},t)$ are solutions to source-free Maxwell equations:

$$\nabla \times \boldsymbol{E}(\boldsymbol{r},t) + \mu \frac{\partial \boldsymbol{H}(\boldsymbol{r},t)}{\partial t} = 0 , \tag{5}$$

$$\nabla \cdot \left(\mu \boldsymbol{H}(\boldsymbol{r},t)\right) = 0 , \tag{6}$$

$$\nabla \times \boldsymbol{H}(\boldsymbol{r},t) - \varepsilon_0 \varepsilon(\boldsymbol{r}) \frac{\partial \boldsymbol{E}(\boldsymbol{r},t)}{\partial t} = 0 , \tag{7}$$

$$\nabla \cdot \left(\varepsilon_0 \varepsilon(\boldsymbol{r}) \boldsymbol{E}(\boldsymbol{r},t)\right) = 0 , \tag{8}$$

where μ is magnetic permeability, ε is relative permittivity (dielectric constant) of the medium, and ε_0 is the free-space permittivity. For non-magnetic materials, which commonly constitute an optical waveguide, the magnetic permeability μ is equal to the free-space permeability μ_0. Assuming a waveguide consisting of linear isotropic media, we can reduce Maxwell's (Eqs. 5–8) to the vector wave equations:

$$\Delta \boldsymbol{E}(\boldsymbol{r},t) - \varepsilon_0\varepsilon(\boldsymbol{r})\mu_0 \frac{\partial^2 \boldsymbol{E}(\boldsymbol{r},t)}{\partial t^2} = \nabla\left(\boldsymbol{E}(\boldsymbol{r},t)\cdot\nabla \ln \varepsilon_0\varepsilon(\boldsymbol{r})\right), \tag{9}$$

$$\Delta \boldsymbol{H}(\boldsymbol{r},t) - \varepsilon_0\varepsilon(\boldsymbol{r})\mu_0 \frac{\partial^2 \boldsymbol{H}(\boldsymbol{r},t)}{\partial t^2} = (\nabla\times\boldsymbol{H})\times\left(\nabla \ln \varepsilon_0\varepsilon(\boldsymbol{r})\right), \tag{10}$$

where the vector differential operators ∇ and Δ are defined as follows:

$$\nabla f = \frac{\delta f}{\delta x}\boldsymbol{x}_0 + \frac{\delta f}{\delta y}\boldsymbol{y}_0 + \frac{\delta f}{\delta z}\boldsymbol{z}_0, \tag{11}$$

$$\nabla\cdot\boldsymbol{A} = \frac{\delta A_x}{\delta x} + \frac{\delta A_y}{\delta y} + \frac{\delta A_z}{\delta z}, \tag{12}$$

$$\nabla\times\boldsymbol{A} = \left(\frac{\delta A_y}{\delta z} - \frac{\delta A_z}{\delta y}\right)\boldsymbol{x}_0 + \left(\frac{\delta A_z}{\delta x} - \frac{\delta A_x}{\delta z}\right)\boldsymbol{y}_0 + \left(\frac{\delta A_x}{\delta y} - \frac{\delta A_y}{\delta x}\right)\boldsymbol{z}_0, \tag{13}$$

$$\Delta\boldsymbol{A} = \left(\frac{\partial^2 A_x}{\partial x^2} + \frac{\partial^2 A_x}{\partial y^2} + \frac{\partial^2 A_x}{\partial z^2}\right)\boldsymbol{x}_0 + \left(\frac{\partial^2 A_y}{\partial x^2} + \frac{\partial^2 A_y}{\partial y^2} + \frac{\partial^2 A_y}{\partial z^2}\right)\boldsymbol{y}_0 + \left(\frac{\partial^2 A_z}{\partial x^2} + \frac{\partial^2 A_z}{\partial y^2} + \frac{\partial^2 A_z}{\partial z^2}\right)\boldsymbol{z}_0, \tag{14}$$

and f and $\boldsymbol{A} = (A_x, A_y, A_z)$ are scalar and vector functions on cartesian coordinates (x, y, z) and $\boldsymbol{x}_0$, $\boldsymbol{y}_0$ and $\boldsymbol{z}_0$ are unit vectors. If we assume translational invariance of the waveguide in the z-direction, propagation along the z-direction, and time dependence of the field vectors in the form of $\exp(-\mathrm{i}\omega t)$, where ω is the angular frequency and $\mathrm{i} = \sqrt{-1}$, the modal fields can be expressed in the separable form:

$$\boldsymbol{E} = \boldsymbol{e}(x,y)\exp\left(\mathrm{i}(\beta z - \omega t)\right) = \left\{\boldsymbol{e}_\mathrm{t}(x,y) + e_z(x,y)\boldsymbol{z}_0\right\}\exp\left(\mathrm{i}(\beta z - \omega t)\right), \tag{15}$$

$$\boldsymbol{H} = \boldsymbol{h}(x,y)\exp\left(\mathrm{i}(\beta z - \omega t)\right) = \left\{\boldsymbol{h}_\mathrm{t}(x,y) + h_z(x,y)\boldsymbol{z}_0\right\}\exp\left(\mathrm{i}(\beta z - \omega t)\right), \tag{16}$$

where β denotes the propagation constant of a mode and subscript t denotes the transversal component of field vectors. For the modal fields described by Eqs. 15 and 16, the vector wave equations can be reduced to:

$$\left\{\Delta_\mathrm{t} + \omega^2\varepsilon\varepsilon_0\mu_0 - \beta^2\right\}\boldsymbol{e} = -\left\{\nabla_t + \mathrm{i}\beta\boldsymbol{z}\right\}\left\{\boldsymbol{e}_\mathrm{t}\nabla_\mathrm{t}\ln\varepsilon\varepsilon_0\right\}, \tag{17}$$

$$\left\{\Delta_\mathrm{t} + \omega^2\varepsilon\varepsilon_0\mu_0 - \beta^2\right\}\boldsymbol{h} = -\left(\nabla_\mathrm{t}\ln\varepsilon\varepsilon_0\right)\times\left(\left\{\nabla_\mathrm{t} + \mathrm{i}\beta\boldsymbol{z}\right\}\times\boldsymbol{h}\right). \tag{18}$$

These vector wave equations are a restatement of Maxwell's equations for an arbitrary refractive index profile. Subject to the requirements that the modal fields are bounded everywhere and decay sufficiently fast at large distances from the waveguide, these equations contain all of the information necessary to determine the modal fields and propagation constants of all the guided modes of the waveguide.

Let us consider an optical waveguide consisting of three homogeneous media (Fig. 1) with a permittivity profile:

$$\varepsilon(x) = \varepsilon_3 = n_3^2\,, \quad x > d\,, \tag{19}$$
$$\varepsilon(x) = \varepsilon_2 = n_2^2\,, \quad -d \le x \le d\,, \tag{20}$$
$$\varepsilon(x) = \varepsilon_1 = n_1^2\,, \quad x < -d\,, \tag{21}$$

where d is the waveguiding layer half-width and ε_i and n_i ($i = 1, 2, 3$) are generally complex permittivities and refractive indices (hereafter, we shall be using ε for the *relative* permittivity unless stated otherwise).

By orienting cartesian axes as shown in Fig. 1, the field vectors depend on x and z only and Eqs. 15 and 16 can be written as:

$$\boldsymbol{E} = \boldsymbol{e}(x) \exp\left(\mathrm{i}(\beta z - \omega t)\right)\,, \tag{22}$$
$$\boldsymbol{H} = \boldsymbol{h}(x) \exp\left(\mathrm{i}(\beta z - \omega t)\right)\,, \tag{23}$$

In each medium the $\nabla_t \ln \varepsilon$ term vanishes and each cartesian field component satisfies a simplified wave equation:

$$\left\{\Delta_t + \omega^2 \varepsilon \varepsilon_0 \mu_0 - \beta^2\right\} e_i = 0\,, \tag{24}$$
$$\left\{\Delta_t + \omega^2 \varepsilon \varepsilon_0 \mu_0 - \beta^2\right\} h_i = 0\,, \tag{25}$$

where $i = x, y, z$.

The solution of Eqs. 24 and 25 yields two linearly independent sets of modes. One set with $h_z = 0$ everywhere, referred as to transverse magnetic (TM); the other with $e_z = 0$ everywhere, referred as to transverse electric (TE). Substitution of the field profiles Eqs. 22 and 23 into Eqs. 24 and 25, respec-

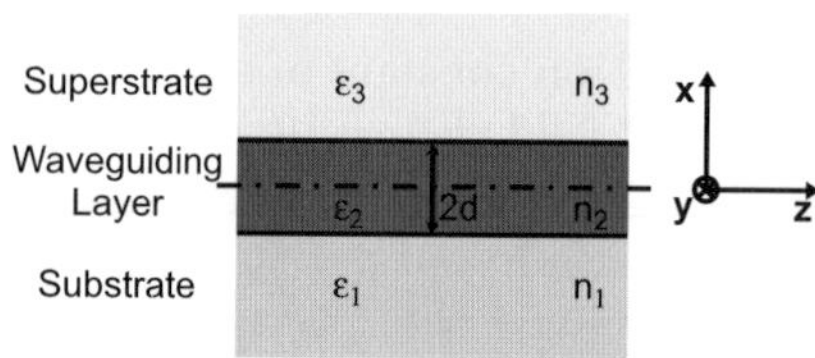

Fig. 1 Section of a planar waveguide with a step refractive index profile

tively, yields for the transversal components of the field vectors:

$$\frac{\partial^2 e_y(x)}{\partial x^2} + (\omega^2 \varepsilon \varepsilon_0 \mu_0 - \beta^2) e_y(x) = 0 ; \quad \text{for the TE modes ,} \tag{26}$$

$$\frac{\partial^2 h_y(x)}{\partial x^2} + (\omega^2 \varepsilon \varepsilon_0 \mu_0 - \beta^2) h_y(x) = 0 ; \quad \text{for the TM modes .} \tag{27}$$

In each medium the solution of wave Eqs. 26 and 27 can be expressed as a linear combination of functions: $\exp(\mathrm{i}\kappa_i x)$ and $\exp(-\mathrm{i}\kappa_i x)$, where $\kappa_i^2 = \omega^2 \varepsilon_i \varepsilon_0 \mu_0 - \beta^2$ $(i = 1, 2, 3)$. The other non-zero components of the field vectors can be determined from Eqs. 5 and 7. This yields:

TE modes:

$$e_y(x) = a_i^+ \exp(\mathrm{i}\kappa_i x) + a_i^- \exp(-\mathrm{i}\kappa_i x) , \tag{28}$$

$$h_x(x) = \frac{\beta}{\mu_0 \omega} \left[a_i^+ \exp(\mathrm{i}\kappa_i x) + a_i^- \exp(-\mathrm{i}\kappa_i x) \right] , \tag{29}$$

$$h_z(x) = -\frac{\kappa_i}{\mu_0 \omega} \left[a_i^+ \exp(\mathrm{i}\kappa_i x) - a_i^- \exp(-\mathrm{i}\kappa_i x) \right] , \text{ and} \tag{30}$$

TM modes:

$$h_y(x) = b_i^+ \exp(\mathrm{i}\kappa_i x) + b_i^- \exp(-\mathrm{i}\kappa_i x) , \tag{31}$$

$$e_x(x) = -\frac{\beta}{\varepsilon_i \varepsilon_0 \omega} \left[b_i^+ \exp(\mathrm{i}\kappa_i x) + b_i^- \exp(-\mathrm{i}\kappa_i x) \right] , \tag{32}$$

$$e_z(x) = \frac{\kappa_i}{\varepsilon_i \varepsilon_0 \omega} \left[b_i^+ \exp(\mathrm{i}\kappa_i x) - b_i^- \exp(-\mathrm{i}\kappa_i x) \right] . \tag{33}$$

Outside the waveguiding layer, modal fields bound to the waveguide are described by only one of these solutions and decay exponentially with an increasing distance from the waveguide. Consequently, in each pair of amplitudes a_1^+ and a_1^- and a_3^+ and a_3^-, one amplitude is equal to zero for TE modes, and in each pair of amplitudes b_1^+ and b_1^- and b_3^+ and b_3^-, one amplitude is equal to zero for TM modes. The boundary conditions of Maxwell's equations require that the components of the electric and magnetic field intensity vectors parallel to the boundaries of the waveguiding layer are continuous at the boundaries ($x = d$ and $x = -d$). These boundary conditions present a homogenous series of four linear equations for four unknown amplitudes, which yields a non-zero solution only if the determinant of the matrix of coefficients is equal to zero. This requirement leads to the eigenvalue equations:

$$\tan(\kappa d) = \frac{\gamma_1/\kappa + \gamma_3/\kappa}{1 - (\gamma_1/\kappa)(\gamma_3/\kappa)} ; \text{ for the TE modes ,} \tag{34}$$

$$\tan(\kappa d) = \frac{\gamma_1 \varepsilon_2/\kappa\varepsilon_1 + \gamma_3 \varepsilon_2/\kappa\varepsilon_3}{1 - (\gamma_1 \varepsilon_2/\kappa\varepsilon_1)(\gamma_3 \varepsilon_2/\kappa\varepsilon_3)} ; \text{ for the TM modes ,} \tag{35}$$

where $\kappa^2 = \omega^2 \varepsilon_2 \varepsilon_0 \mu_0 - \beta^2$ and $\gamma_{1,3}^2 = \beta^2 - \omega^2 \varepsilon_{1,3} \varepsilon_0 \mu_0$.

The eigenvalue Eqs. 34 and 35 are transcendental equations for unknown modal propagation constants. After solving the eigenvalue equations, the field profiles can be determined by substituting the values of modal propagation constants β into the boundary conditions and calculating the amplitudes a_i^+ and a_i^- for TE modes and b_i^+ and b_i^- for TM modes ($i = 1, 2, 3$).

If the media constituting the waveguide are lossless ($\varepsilon_1, \varepsilon_2$, and ε_3 are real positive numbers), the propagation constants are also real. Propagation constants of modes of a waveguide containing absorbing media (e.g., metal) are complex. The propagation constant is related to the modal effective index n_{ef} and modal attenuation b as follows:

$$n_{ef} = \frac{c}{\omega} \operatorname{Re}\{\beta\} , \tag{36}$$

$$b = \operatorname{Im}\{\beta\} \frac{0.2}{\ln 10} , \tag{37}$$

where Re{} and Im{} denote the real and imaginary parts of a complex number, respectively, and c denotes the speed of light in vacuum; the modal attenuation b is in dB cm^{-1} if β is given in m^{-1}.

2.1
Surface Plasmons on Metal–Dielectric Waveguides

A waveguide consisting of a semi-infinite metal with a complex permittivity $\varepsilon_m = \varepsilon'_m + i\varepsilon''_m$, and a semi-infinite dielectric with permittivity $\varepsilon_d = \varepsilon'_d + i\varepsilon''_d$, where ε'_i and ε''_i are real and imaginary parts of ε_i (i is m or d), see Fig. 2, can be treated as a limiting case of a three-layer waveguide (Fig. 1) with a metal substrate, a dielectric superstrate, and a waveguiding layer with a thickness equal to zero.

The propagation constants of the guided modes propagating along such a structure are the solutions of Eqs. 34 and 35, which for $d = 0$ can be rewritten as:

$$\gamma_m = - \gamma_d ; \quad \text{for the TE modes} , \tag{38}$$

$$\frac{\gamma_m}{\varepsilon_m} = - \frac{\gamma_d}{\varepsilon_d} ; \quad \text{for the TM modes} , \tag{39}$$

where $\gamma_i^2 = \beta^2 - \omega^2 \mu_0 \varepsilon_0 \varepsilon_i$ (i is m or d). The eigenvalue equation for TE modes (1.38) yields no solution that would represent a bounded mode. The TM

Fig. 2 A metal–dielectric waveguide

mode eigenvalue (Eq. 39) can be reduced to:

$$\beta = \frac{\omega}{c}\sqrt{\frac{\varepsilon_d \varepsilon_m}{\varepsilon_d + \varepsilon_m}} = k\sqrt{\frac{\varepsilon_d \varepsilon_m}{\varepsilon_d + \varepsilon_m}}\,, \tag{40}$$

where c is the speed of light in vacuum and $k = 2\pi/\lambda$ is the free-space wavenumber, where λ is the free-space wavelength [6, 7]. For lossless metal and dielectric ($\varepsilon''_m = \varepsilon''_d = 0$), Eqs. 39 and 40 represent a guided mode, providing that the permittivities ε'_m and ε'_d are of opposite signs, and that $\varepsilon'_m < -\varepsilon'_d$. This guided mode is sometimes referred as to the Fano mode [7]. As the permittivity of dielectric materials is usually positive, for the Fano mode to exist, the real part of the permittivity of the metal needs to be negative. For metals following the free-electron model [13]:

$$\varepsilon_m = \varepsilon_0 \left(1 - \frac{\omega_p^2}{\omega^2 + i\omega\nu}\right), \tag{41}$$

where ν is the collision frequency and ω_p is the plasma frequency:

$$\omega_p = \sqrt{\frac{Ne^2}{\varepsilon_0 m_e}}\,, \tag{42}$$

where N is the concentration of free electrons, and e and m_e are the electron charge and mass, respectively, this requirement is fulfilled for frequencies lower than the plasma frequency of the metal. As shown in Fig. 3 metals such as gold, silver and aluminum exhibit a negative real part of permittivity in visible and near infrared region of the spectrum.

Absorption, which in reality always exists, introduces a non-zero imaginary part into the permittivity of metals (Fig. 3, lower plot) and permits the existence of guided modes even for $\varepsilon'_m > -\varepsilon_d$. These modes, sometimes referred as to evanescent modes [7], exhibit a very high attenuation and are therefore less practically important. In this work, we shall refer to all of the guided modes described by eigenvalue (Eq. 40) as *surface plasmons* (SP).

If the real part of the permittivity of the metal is negative and its magnitude is much larger than the imaginary part $|\varepsilon'_m| \gg \varepsilon''_m$, the complex propagation constant of the surface plasmon given by Eq. 40 can be expressed as:

$$\beta = \beta' + i\beta'' \doteq \frac{\omega}{c}\sqrt{\frac{\varepsilon'_m \varepsilon_d}{\varepsilon'_m + \varepsilon_d}} + i\frac{\varepsilon''_m}{2(\varepsilon'_m)^2}\frac{\omega}{c}\left(\frac{\varepsilon'_m \varepsilon_d}{\varepsilon'_m + \varepsilon_d}\right)^{3/2}, \tag{43}$$

where β' and β'' denote the real and imaginary parts of the propagation constant β [6]. As follows from Eq. 43, the imaginary part of the permittivity of metal ε''_m causes the propagation constant of a surface plasmon to have a non-zero imaginary part, which is associated with attenuation of the surface plasmon. The attenuation is sometimes characterized by the propagation length L, which is defined as the distance in the direction of propagation at

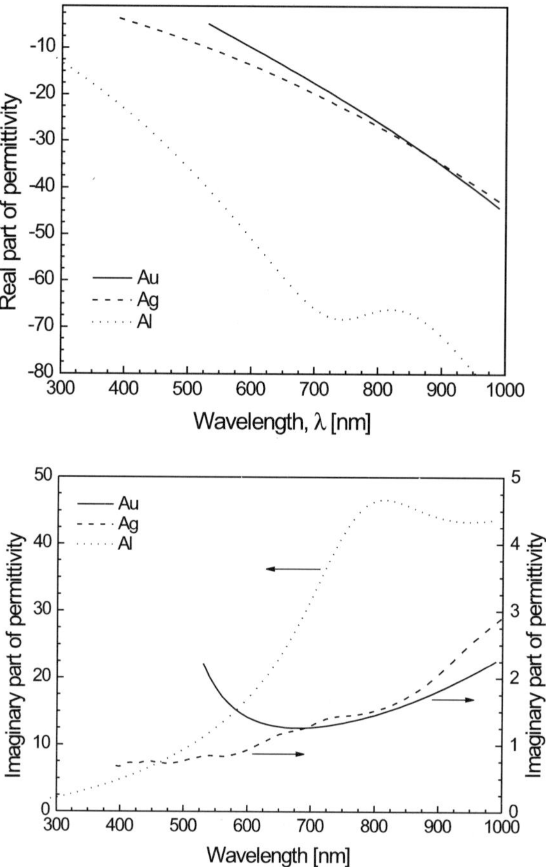

Fig. 3 Permittivity of gold, silver and aluminum as a function of wavelength. Real part of permittivity (*upper plot*) and imaginary part of permittivity (*lower plot*). Data determined ellipsometrically or taken from [14]

which the energy of the surface plasmon decreases by a factor of 1/e:

$$L = 1/\left[2\beta''\right] . \tag{44}$$

Spectral dependencies of the effective index, attenuation, and propagation length of a surface plasmon supported by gold, silver and aluminum are shown in Fig. 4.

As follows from Fig. 4, the existence of a surface plasmon on a metal–dielectric interface is confined to wavelengths longer than a certain critical wavelength, which depends on the plasma frequency and is specific to the metal. For metals such as gold, silver, and aluminum this critical wavelength lies in the UV or visible region. The effective index of a surface plasmon is larger than the effective index of a light wave in the dielectric medium and decreases with increasing wavelength. Attenuation of a surface plasmon

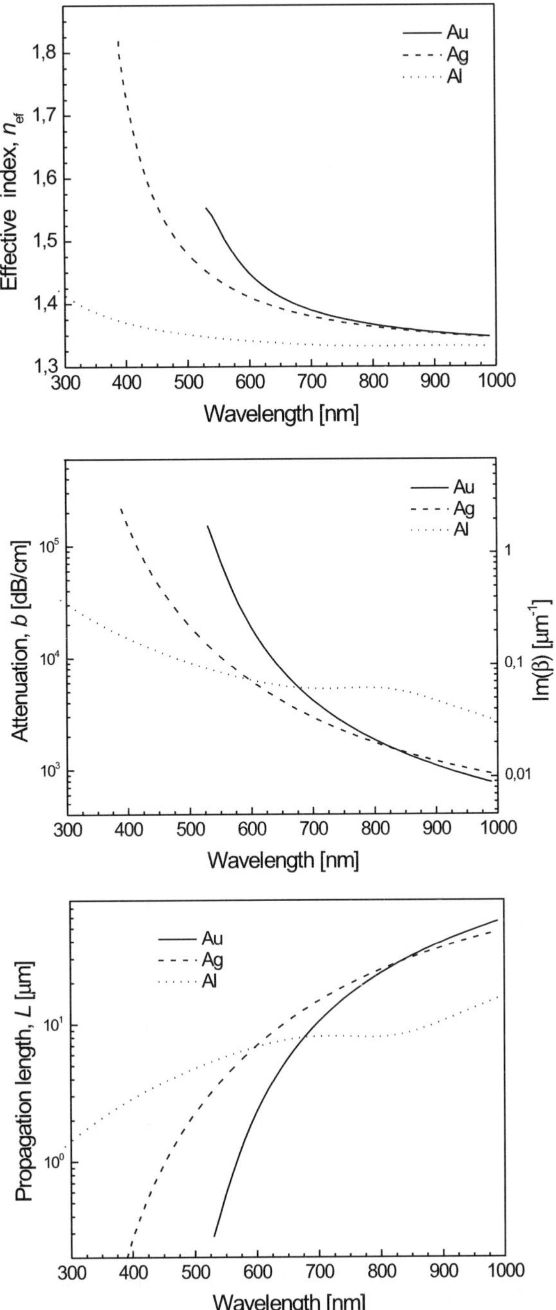

Fig. 4 Effective index, attenuation and propagation length of a surface plasmon propagating along the interface between a dielectric (refractive index 1.32) and a metal as a function of wavelength calculated for gold (Au), silver (Ag), and aluminum (Al)

follows the same trend. As the attenuation of a surface plasmon is proportional to $\varepsilon''_{\mathrm{m}}/\varepsilon'^{2}_{\mathrm{m}}$, the effect of the imaginary part of the permittivity of the metal can be outweighed by the real part of the permittivity. For instance, aluminum exhibits a much larger imaginary part of permittivity than silver. However, the surface plasmons on silver and aluminum suffer approximately the same attenuation at a wavelength of 600 nm as the real part of the permittivity of aluminum is much larger than that of silver. In the wavelength range 550–1000 nm, typical propagation lengths of surface plasmons are 0.6–50 μm, 4–50 μm, and 6–14 μm, for gold, silver and aluminum, respectively.

The distribution of electric and magnetic intensity vectors of a surface plasmon can be obtained from Eqs. 31–33:

$$h_y(x) = A\exp(\gamma_{\mathrm{m}}x) \quad \text{for} \quad x<0 \quad \text{and}$$
$$h_y(x) = A\exp(-\gamma_{\mathrm{d}}x) \quad \text{for} \quad x>0 \tag{45}$$

$$e_x(x) = A\frac{\beta}{\omega\varepsilon_{\mathrm{m}}\varepsilon_0}\exp(\gamma_{\mathrm{m}}x) \quad \text{for} \quad x<0 \quad \text{and}$$
$$e_x(x) = A\frac{\beta}{\omega\varepsilon_{\mathrm{d}}\varepsilon_0}\exp(-\gamma_{\mathrm{d}}x) \quad \text{for} \quad x>0 \tag{46}$$

$$e_z(x) = A\frac{\gamma_{\mathrm{m}}}{\omega\varepsilon_{\mathrm{m}}\varepsilon_0}\exp(\gamma_{\mathrm{m}}x) \quad \text{for} \quad x<0 \quad \text{and}$$
$$e_z(x) = -A\frac{\gamma_{\mathrm{d}}}{\omega\varepsilon_{\mathrm{d}}\varepsilon_0}\exp(-\gamma_{\mathrm{d}}x) \quad \text{for} \quad x>0\,, \tag{47}$$

where:

$$\gamma_{\mathrm{m}} = ik\frac{\varepsilon_{\mathrm{m}}}{\sqrt{\varepsilon_{\mathrm{m}}+\varepsilon_{\mathrm{d}}}} \quad \text{and} \quad \gamma_{\mathrm{d}} = ik\frac{\varepsilon_{\mathrm{d}}}{\sqrt{\varepsilon_{\mathrm{m}}+\varepsilon_{\mathrm{d}}}}\,, \tag{48}$$

and the signs of the square roots in Eq. 48 are chosen so that the real parts of γ_{m} and γ_{d} are positive. A denotes the modal field amplitude.

As follows from Fig. 5, the electromagnetic field of a surface plasmon reaches its maximum at the metal–dielectric interface and decays into both media. The field decay in the direction perpendicular to the metal–dielectric interface is characterized by the penetration depth L_{p}, which is defined as the distance from the interface at which the amplitude of the field decreases by a factor of 1/e:

$$L_{\mathrm{pm}} = 1/\operatorname{Re}\{\gamma_{\mathrm{m}}\} \quad \text{and} \quad L_{\mathrm{pd}} = 1/\operatorname{Re}\{\gamma_{\mathrm{d}}\} \tag{49}$$

The spectral dependence of the penetration depth of a surface plasmon at the interface between gold and a non-dispersive medium with a refractive index of 1.32 is shown in Fig. 6. As follows from Fig. 6, with an increasing wavelength, the portion of the electromagnetic field carried in the dielectric increases and the field of the surface plasmon extends farther into the dielectric.

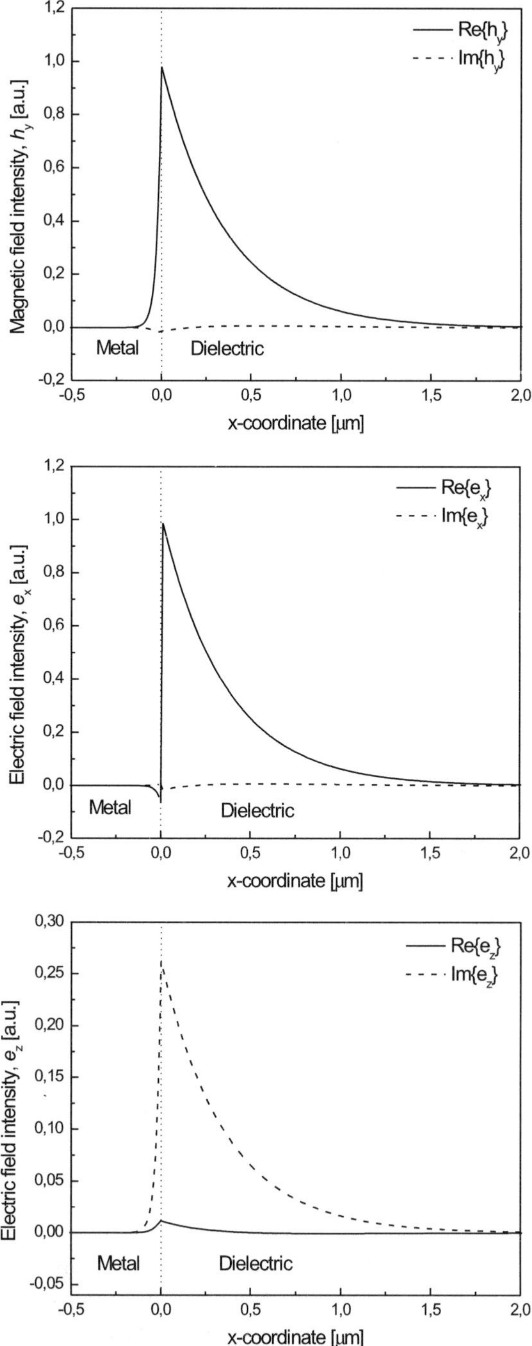

Fig. 5 Distribution of electric and magnetic field of a surface plasmon at the interface of gold ($\varepsilon_m = -25 + 1.44i$) and dielectric (refractive index 1.32), wavelength 800 nm

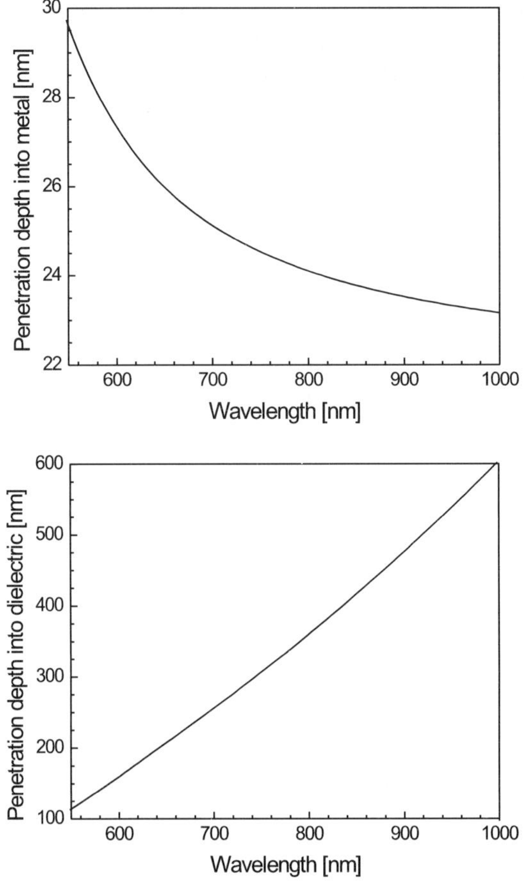

Fig. 6 Penetration depth of a surface plasmon into the metal (*upper plot*) and dielectric (*lower plot*) as a function of wavelength for a surface plasmon propagating along the interface of gold and a dielectric (refractive index 1.32)

2.2 Surface Plasmons on Dielectric–Metal–Dielectric Waveguides

Another example of a planar waveguide supporting surface plasmons is a thin metal film sandwiched between two semi-infinite dielectric media (Fig. 7). If the metal film is much thicker than the penetration depth of a surface plasmon at each metal–dielectric interface, this waveguide supports two TM modes, which correspond to two surface plasmons at the opposite boundaries of the metal film. When the metal thickness decreases, coupling between the two surface plasmons occurs, giving rise to mixed modes of electromagnetic field.

The modes of a dielectric–metal–dielectric waveguide can be found by solving the eigenvalue (Eq. 35). Numerical solutions of this eigenvalue equation for a symmetric waveguide structure ($n_{d1} = n_{d2}$) are shown in Fig. 8. For any thickness of the metal film, there are two coupled surface plasmons, which are referred as to the symmetric and antisymmetric surface plasmons,

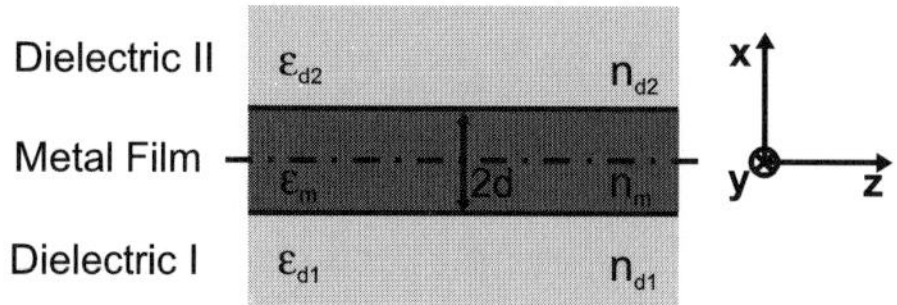

Fig. 7 Thin metal layer sandwiched between two dielectrics

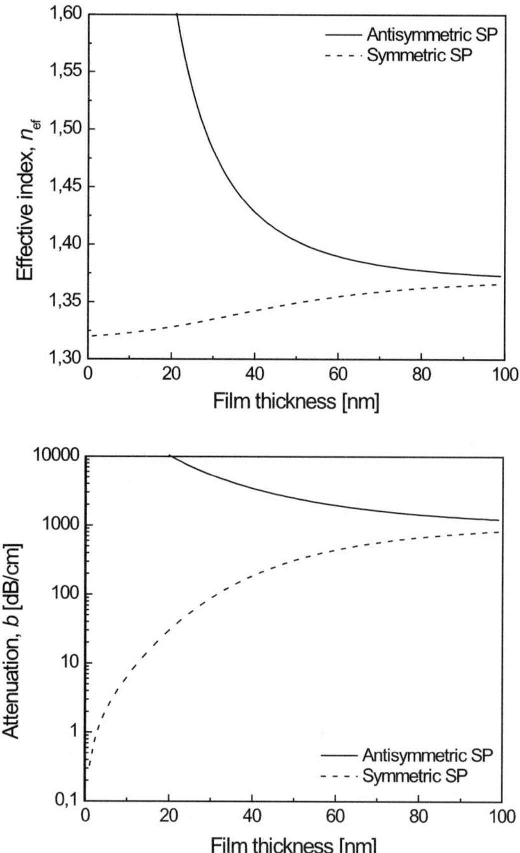

Fig. 8 Effective index and modal attenuation of surface plasmons propagating along a thin gold film ($\varepsilon_m = -25 + 1.44i$) sandwiched between two identical dielectrics ($n_{d1} = n_{d2} = 1.32$) as a function of the thickness of the gold film; wavelength 800 nm

based on the symmetry of the magnetic intensity distribution [14, 15]. The symmetric surface plasmon exhibits effective index and attenuation, which both increase with an increasing metal film thickness. The effective index and attenuation of the antisymmetric surface plasmon decrease with an increasing thickness of the metal film. If the waveguide is asymmetric, the effective index of the symmetric surface plasmon decreases with a decreasing metal film thickness and at a certain metal film thickness, the symmetric surface plasmon ceases to exist as a guided mode, Fig. 9, (this phenomenon is referred as to the mode cut-off). The symmetric surface plasmon exhibits a lower attenuation than its antisymmetric counterpart and therefore it is sometimes referred as to a long-range surface plasmon [16], while the antisymmetric mode is referred as to a short-range surface plasmon [14, 15].

Figures 10 and 11 show the field vector profiles of the symmetric and antisymmetric surface plasmons on a thin gold film surrounded by two identical dielectrics. The profiles of magnetic intensity h_y of symmetric and antisymmetric plasmons are symmetric or antisymmetric with respect to the center

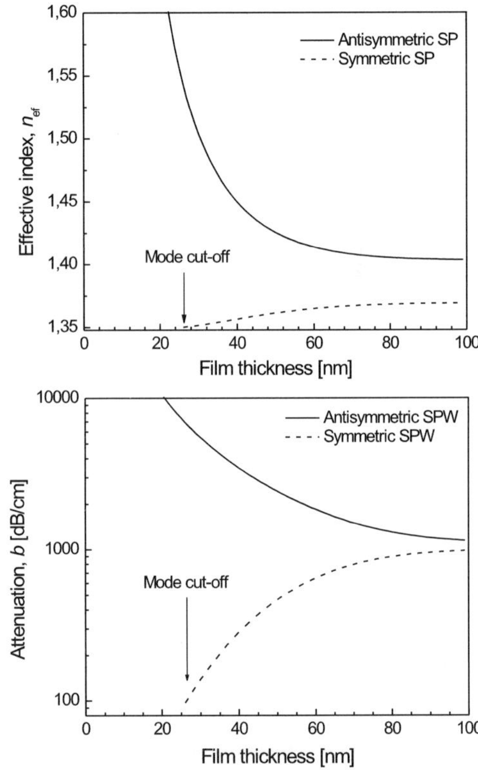

Fig. 9 Effective index and modal attenuation of surface plasmons propagating along a thin gold film ($\varepsilon_m = -25 + 1.44i$) sandwiched between two dielectrics ($n_{d1} = 1.32$ and $n_{d2} = 1.35$) as a function of the thickness of the gold film; wavelength 800 nm

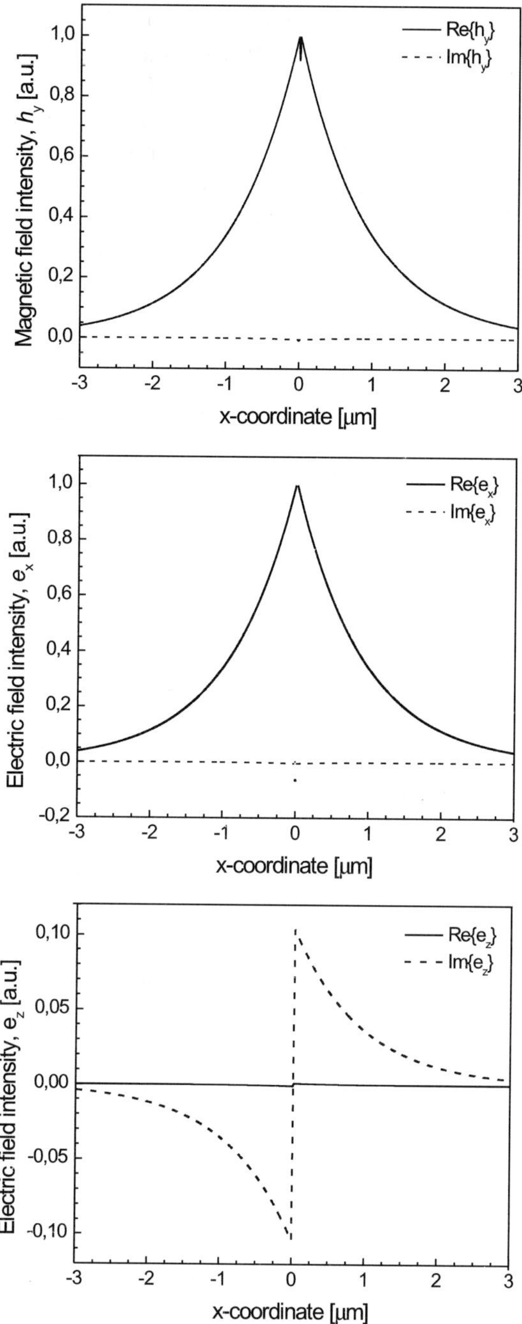

Fig. 10 Field profile of a symmetric surface plasmon on a thin gold film ($\varepsilon_m = -25 + 1.44i$) sandwiched between two identical dielectrics ($n_{d1} = n_{d2} = 1.32$), thickness of the gold film 20 nm, wavelength 800 nm

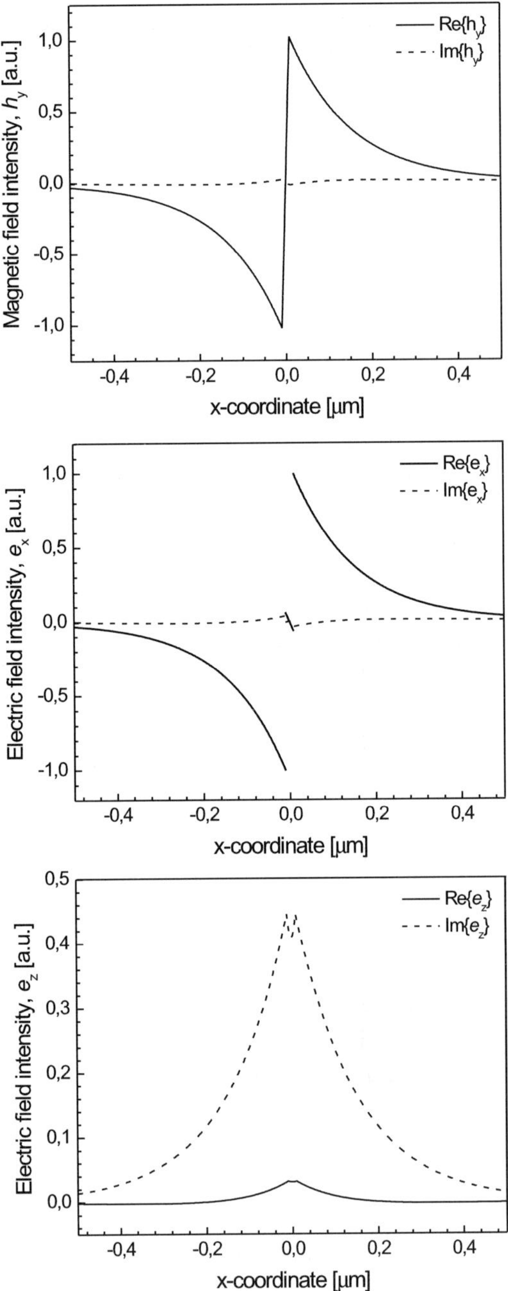

Fig. 11 Field profile of an antisymmetric surface plasmon on a thin gold film ($\varepsilon_{\mathrm{m}} = -25 + 1.44\mathrm{i}$) sandwiched between two identical dielectrics ($n_{\mathrm{d}1} = n_{\mathrm{d}2} = 1.32$), thickness of the gold film 20 nm, wavelength 800 nm

of the metal. The field of the symmetric surface plasmon penetrates deeper into the dielectric media than that of the antisymmetric surface plasmon.

3 Surface Plasmons on Waveguides with a Perturbed Refractive Index Profile

Surface plasmons are characterized by a (complex) propagation constant and a distribution of their electromagnetic field. The propagation constant is a solution of an appropriate eigenvalue equation and depends on the refractive index profile of the waveguide and angular frequency of surface plasmon. If the refractive index profile of the waveguide is perturbed, the propagation constant of the surface plasmon changes. The relationship between the change in the propagation constant of a surface plasmon and a perturbation in the refractive index profile can be analyzed using the perturbation theory [10].

In the perturbation theory, we assume that the magnetic field vector h_y of a surface plasmon supported by a general planar waveguide with and without the refractive index profile perturbation is described by Eq. 18. For the unperturbed and perturbed waveguide with permittivity profiles $\varepsilon(x)$ and $\bar{\varepsilon}(x) = \varepsilon(x) + \delta\varepsilon(x)$, respectively, this equation can be rewritten as:

$$\left\{\frac{\partial^2}{\partial x^2} + \omega^2 \varepsilon\mu - \beta^2\right\} h_y = \frac{\partial \ln \varepsilon}{\partial x}\frac{\partial}{\partial x} h_y \quad \text{for the unperturbed waveguide, and} \tag{50}$$

$$\left\{\frac{\partial^2}{\partial x^2} + \omega^2 \bar{\varepsilon}\mu - \overline{\beta}^2\right\} \overline{h}_y = \frac{\partial \ln \bar{\varepsilon}}{\partial x}\frac{\partial}{\partial x} \overline{h}_y \quad \text{for the unperturbed waveguide,} \tag{51}$$

where $\overline{\beta}$ and $\overline{h}_y$ denote the perturbed modal propagation constant and modal field, respectively. If we multiply Eq. 50 with $\overline{h}_y/\bar{\varepsilon}$, Eq. 51 with h_y/ε, subtract the two equations, and integrate the resulting equation over the cross-section of the waveguide A_∞, we obtain [17]:

$$\beta^2 - \overline{\beta}^2 = \frac{\beta^2 \int\limits_{A_\infty} \left(\frac{1}{\bar{\varepsilon}} - \frac{1}{\varepsilon}\right) h_y \bar{h}_y \, \mathrm{d}A + \int\limits_{A_\infty} \left(\frac{1}{\bar{\varepsilon}} - \frac{1}{\varepsilon}\right) \frac{\partial h_y}{\partial x}\frac{\partial \bar{h}_y}{\partial x} \, \mathrm{d}A}{\int\limits_{A_\infty} \frac{1}{\bar{\varepsilon}} h_y \bar{h}_y \, \mathrm{d}A} . \tag{52}$$

For a small permittivity profile perturbation $|\delta\varepsilon(x)| \ll |\varepsilon(x)|$, we can assume that the modal field remains unchanged ($h_y \doteq \bar{h}_y$) and the modal propagation constant is altered only slightly $\left(\left|\beta - \overline{\beta}\right| \ll |\beta|\right)$. Then, Eq. 52 can be reduced to:

$$\delta\beta = \frac{\beta^2 \int\limits_{A_\infty} \frac{\delta\varepsilon}{\varepsilon^2} h_y^2 \, dA + \int\limits_{A_\infty} \frac{\delta\varepsilon}{\varepsilon^2} \left(\frac{\partial h_y}{\partial x}\right)^2 dA}{2\beta \int\limits_{A_\infty} \frac{1}{\varepsilon} h_y^2 \, dA} . \tag{53}$$

Furthermore, we shall apply this perturbation formula to the investigation of the effect of selected types of refractive index changes on (a) surface plasmons propagating along a single metal–dielectric interface (metal–dielectric waveguide) and (b) coupled surface plasmons propagating along a thin metal film (dielectric–metal–dielectric waveguide), Fig. 12.

Two main types of refractive index perturbations will be discussed here in detail. The first type is a homogeneous change in the refractive index in the whole superstrate, Fig. 13, (herein referred as to bulk refractive index change), which can be described by a change in the permittivity profile, $\varepsilon(x) \to \bar{\varepsilon}(x)$, where:

$$\varepsilon(x) = \begin{cases} \varepsilon_\mathrm{d} \\ \varepsilon_\mathrm{m} \end{cases} \quad \text{and} \quad \bar{\varepsilon}(x) = \begin{cases} \varepsilon_\mathrm{d} + \delta\varepsilon \\ \varepsilon_\mathrm{m} \end{cases} \quad \text{for} \quad \begin{matrix} x > 0 \\ x \leq 0 \end{matrix} . \tag{54}$$

The second type of perturbation is a homogenous change in the refractive index that occurs within a limited distance h from the surface of the metal film which is smaller than the penetration depth of a surface plasmon, Fig. 14, (herein referred as to surface refractive index change). Such a refractive index perturbation is characterized by a permittivity profile change $\varepsilon(x) \to \bar{\varepsilon}(x)$, where:

$$\varepsilon(x) = \begin{cases} \varepsilon_\mathrm{d} \\ \varepsilon_\mathrm{d} \\ \varepsilon_\mathrm{m} \end{cases} \quad \text{and} \quad \bar{\varepsilon}(x) = \begin{cases} \varepsilon_\mathrm{d} \\ \varepsilon_\mathrm{d} + \delta\varepsilon \\ \varepsilon_\mathrm{m} \end{cases} \quad \text{for} \quad \begin{matrix} x \geq h \\ 0 < x < h \\ x \leq 0 \end{matrix} , \tag{55}$$

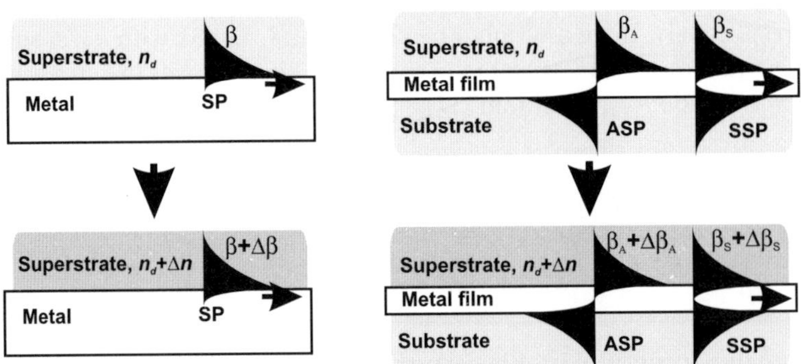

Fig. 12 Surface plasmons on a metal–dielectric waveguide (*left*) and a dielectric–metal–dielectric waveguide (*right*) with a perturbed refractive index of superstrate

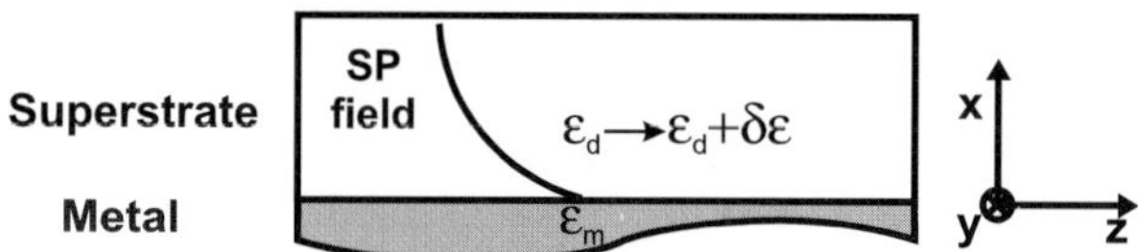

Fig. 13 Refractive index change occurring within a whole superstrate

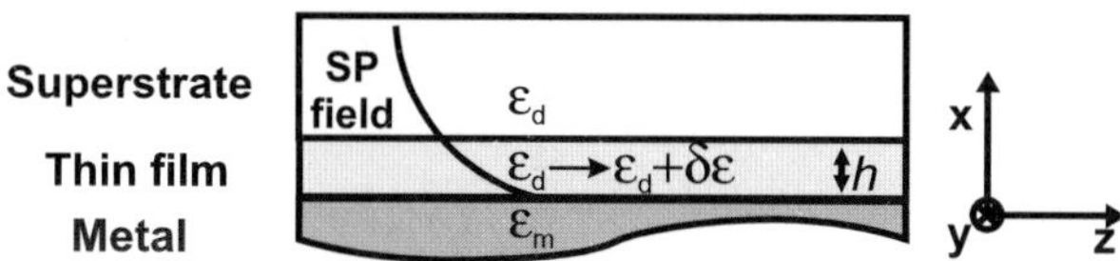

Fig. 14 A homogeneous refractive index change occurring within a short distance from the metal surface

3.1 Perturbed Surface Plasmons on Metal–Dielectric Waveguides

A change in the propagation constant of a surface plasmon on a metal–dielectric interface produced by a bulk refractive index change can be calculated by substituting the perturbation of the permittivity profile (Eq. 54) and the field distribution of the surface plasmon (Eq. 48) into the perturbation formula (Eq. 53). After a straightforward manipulation, the following analytical expressions for the perturbations in the propagation constant $\delta\beta$, and the effective refractive index δn_{ef} can be obtained:

$$\delta\beta = \frac{\beta^3}{2k^2\varepsilon_{\mathrm{d}}^2}\delta\varepsilon = \frac{\beta^3}{k^2 n_{\mathrm{d}}^3}\delta n\,, \tag{56}$$

$$\delta n_{\mathrm{ef}} = \frac{n_{\mathrm{ef}}^3}{n_{\mathrm{d}}^3}\delta n\,. \tag{57}$$

where the perturbation in the refractive index and permittivity are related as $\delta\varepsilon = 2n_{\mathrm{d}}\delta n$. As the effective index of the surface plasmon at a metal–dielectric interface n_{ef} is always larger than the refractive index of the dielectric n_{d}, the bulk refractive index sensitivity of the effective index of the surface plasmon $(\delta n_{\mathrm{ef}}/\delta n)_{\mathrm{B}}$ is always larger than the sensitivity of a free space plane wave in the infinite dielectric medium (which is equal to one). For metals with a negative real part of the permittivity $\varepsilon'_{\mathrm{m}} < 0$ and a magnitude much larger than the imaginary part $|\varepsilon'_{\mathrm{m}}| \gg \varepsilon''_{\mathrm{m}}$, the sensitivity of the effective index of the surface plasmon to a bulk refractive change can be expressed as:

$$\left(\frac{\delta n_{\mathrm{ef}}}{\delta n}\right)_{\mathrm{B}} \doteq \left(\frac{\varepsilon'_{\mathrm{m}}}{\varepsilon'_{\mathrm{m}} + n_{\mathrm{d}}^2}\right)^{3/2}\,. \tag{58}$$

Equation 58 suggests that the sensitivity depends on the real part of the permittivity of the metal and decreases with its increasing magnitude. As the magnitude of the real part of the permittivity of gold decreases with an increasing wavelength (Fig. 3), the dependence of $(\delta n_{\mathrm{ef}}/\delta n)_{\mathrm{B}}$ on the wavelength follows the same trend, Fig. 15. For gold as a surface plasmon-supporting metal, both the results of the perturbation theory (Eq. 57) and its approximation (Eq. 58) agree very well with the rigorous approach based on the numerical calculation of the effective index of surface plasmon for the perturbed and unperturbed waveguide.

A change in the surface plasmon propagation constant induced by a surface refractive index change occurring within a layer with a thickness h can be calculated by substituting the perturbation of the permittivity profile Eq. 55 and the field distribution of the surface plasmon Eq. 48 into Eq. 53. After a straightforward mathematical manipulation we obtain:

$$\delta\beta = \frac{\beta^3}{2k^2\varepsilon_{\mathrm{d}}^2}\left[1 - \exp(-2\gamma_{\mathrm{d}}h)\right]\delta\varepsilon = \frac{\beta^3}{k^2 n_{\mathrm{d}}^3}\left[1 - \exp(-2\gamma_{\mathrm{d}}h)\right]\delta n\,, \tag{59}$$

where $\gamma_{\mathrm{d}} = \sqrt{\beta^2 - \omega^2\mu_0\varepsilon_0\varepsilon_{\mathrm{d}}}$ (a sign of the square root is selected so that $\mathrm{Re}\{\gamma_{\mathrm{d}}\} > 0$). For the perturbation of the effective index of the surface plasmon, this equation yields:

$$\delta n_{\mathrm{ef}} = \frac{\mathrm{Re}\left\{\beta^3\left[1 - \exp(-2\gamma_{\mathrm{d}}h)\right]\right\}}{k^3 n_{\mathrm{d}}^3}\delta n\,. \tag{60}$$

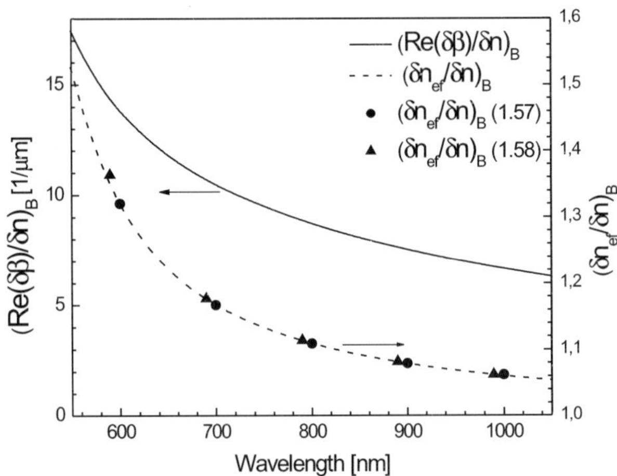

Fig. 15 Sensitivity of the real propagation constant (—) and effective index (- - -) of a surface plasmon on a metal–dielectric interface to a bulk refractive index change as a function of wavelength calculated rigorously from eigenvalue equation and using the perturbation theory. Waveguiding structure: gold–dielectric ($n_{\mathrm{d}} = 1.32$)

The perturbation of the effective index of a surface plasmon depends exponentially on the thickness of the layer within which the refractive index change occurs. For a thicknesses much larger than the penetration depth of the surface plasmon ($h \gg L_{\mathrm{pd}} = 1/\operatorname{Re}\{\gamma_{\mathrm{d}}\}$), the exponential term can be neglected and Eq. 60 simplifies to Eq. 57. For refractive index changes occurring within a layer thinner than the penetration depth of the field of the surface plasmon ($h \ll L_{\mathrm{pd}} = 1/\operatorname{Re}\{\gamma_{\mathrm{d}}\}$), the expressions for the perturbations in the propagation constant and the effective refractive index can be reduced to:

$$\delta\beta = \frac{2\gamma_{\mathrm{d}}\beta^3}{k^2 n_{\mathrm{d}}^3} h\delta n\,, \tag{61}$$

$$\delta n_{\mathrm{ef}} = \frac{2\operatorname{Re}\left\{\gamma_{\mathrm{d}}\beta^3\right\}}{k^3 n_{\mathrm{d}}^3} h\delta n\,. \tag{62}$$

Figure 16 shows the sensitivity of the propagation constant $(\operatorname{Re}(\delta\beta)/\delta n)_{\mathrm{S}}$ and effective index $(\delta n_{\mathrm{ef}}/\delta n)_{\mathrm{S}}$ to a surface refractive index change calculated for a surface plasmon supported on gold and a refractive index change occurring within a 5 nm thick layer at the surface of the metal supporting a surface plasmon. As the layer thickness is much smaller than the penetration depth of the field of the surface plasmon on the considered structure, the sensitivity is a linear function of the thickness of the layer h.

If the real part of the permittivity of the metal is much larger than the imaginary part $\left|\varepsilon'_{\mathrm{m}}\right| \gg \varepsilon''_{\mathrm{m}}$, the sensitivity of the effective index of a surface

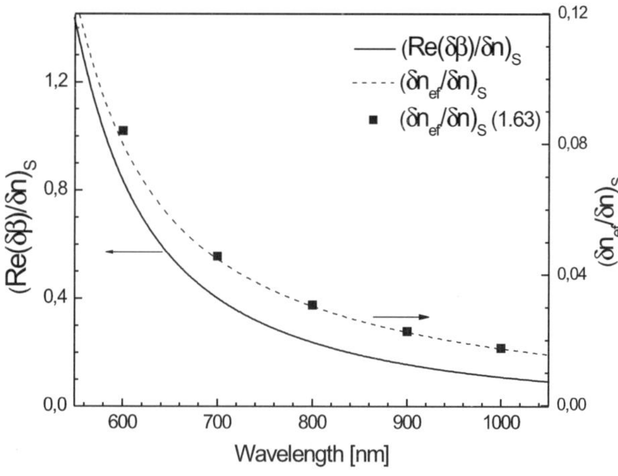

Fig. 16 Sensitivity of the propagation constant and effective index of a surface plasmon on a metal–dielectric interface to a surface refractive index change as a function of wavelength. Waveguiding structure: gold–thin dielectric film ($h = 5$ nm), dielectric superstrate ($n_{\mathrm{d}} = 1.32$)

plasmon to a surface refractive index change can be expressed as:

$$\left(\frac{\delta n_{\mathrm{ef}}}{\delta n}\right)_{\mathrm{S}} = 2\frac{n_{\mathrm{ef}}^3}{n_{\mathrm{d}}^3}\frac{h}{L_{\mathrm{pd}}} = \left(\frac{\delta n_{\mathrm{ef}}}{\delta n}\right)_{\mathrm{B}}\frac{2h}{L_{\mathrm{pd}}}\,. \tag{63}$$

By employing the approximate expressions for the bulk refractive index sensitivity of the effective index Eq. 58 and low-loss metal approximation of L_{pd} (Eq. 48), Eq. 63 can be reduced to:

$$\left(\frac{\delta n_{\mathrm{ef}}}{\delta n}\right)_{\mathrm{S}} \doteq \left(\frac{\varepsilon'_{\mathrm{m}}}{\varepsilon'_{\mathrm{m}} + n_{\mathrm{d}}^2}\right)^{\frac{3}{2}} \frac{2n_{\mathrm{d}}^2}{\sqrt{-\varepsilon'_{\mathrm{m}} - n_{\mathrm{d}}^2}} hk\,. \tag{64}$$

As follows from Eq. 63, the sensitivity of the effective index to a surface refractive index change is proportional to the bulk refractive index sensitivity and the thickness of the layer within which the surface refractive index change occurs, and is inversely proportional to the penetration depth of the surface plasmon. As the penetration depth of a surface plasmon on gold increases with increasing wavelength, the surface refractive index sensitivity of the effective index (Fig. 16) decreases with the wavelength faster than the bulk refractive index sensitivity (Fig. 15). As illustrated in Fig. 16, the approximate equation for the sensitivity to as surface refractive index change (Eq. 63) yields results that are in a good agreement with the rigorous approach based on the numerical calculation of the effective index of surface plasmon for the perturbed and unperturbed waveguide from the appropriate eigenvalue equations.

3.2
Perturbed Surface Plasmons on Dielectric–Metal–Dielectric Waveguides

Perturbation of symmetric and antisymmetric surface plasmons (Sect. 2.2) propagating along a thin metal film with a thickness $2d$ can be calculated by determining the propagation constants of the surface plasmons supported by the unperturbed and perturbed waveguides as solutions to the eigenvalue (Eq. 35). The sensitivity of the effective index $(\delta n_{\mathrm{ef}}/\delta n)_{\mathrm{B}}$ to bulk refractive index changes in the superstrate as a function of metal layer thickness is shown in Fig. 17. For the considered structure and thicknesses of the metal film, the sensitivity of the antisymmetric surface plasmon is higher than that of its symmetric counterpart. The sensitivity of the symmetric surface plasmon increases with the thickness of the metal film, while the sensitivity of the antisymmetric surface plasmon follows an opposite trend. For thick metal films, the coupled surface plasmons consists of two weakly coupled surface plasmons propagating on opposite surfaces of the metal film and therefore the sensitivities of both the symmetric and antisymmetric surface plasmons approach the value of one half of the sensitivity of the surface plasmon at a single metal–dielectric interface.

Figure 18 shows the sensitivity of the effective index $(\delta n_{\mathrm{ef}}/\delta n)_{\mathrm{B}}$ to bulk refractive index changes for symmetric and antisymmetric surface plasmons on a thin gold film. While the sensitivity of the effective index of the antisymmetric surface plasmon decreases with an increasing wavelength, sensitivity of its symmetric counterpart varies only slightly over the considered wavelength range.

The sensitivity of the effective index of the symmetric and antisymmetric surface plasmons to a surface refractive index change, Figs. 19 and 20, follows basically the same trends as the sensitivity to bulk refractive index changes.

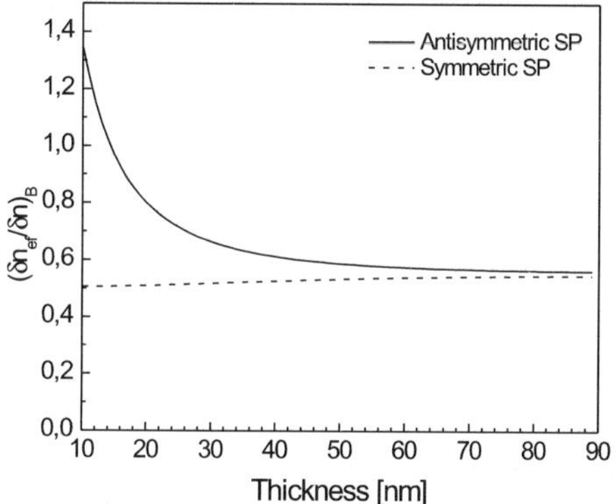

Fig. 17 Sensitivity of the effective index of symmetric and antisymmetric surface plasmons to bulk refractive index changes as a function of the thickness of metal layer. Waveguide configuration: dielectric ($n_1 = 1.32$)–gold ($\varepsilon_{\mathrm{m}} = -25 + 1.44\mathrm{i}$)–dielectric superstrate ($n_{\mathrm{d}} = 1.32$), wavelength 800 nm

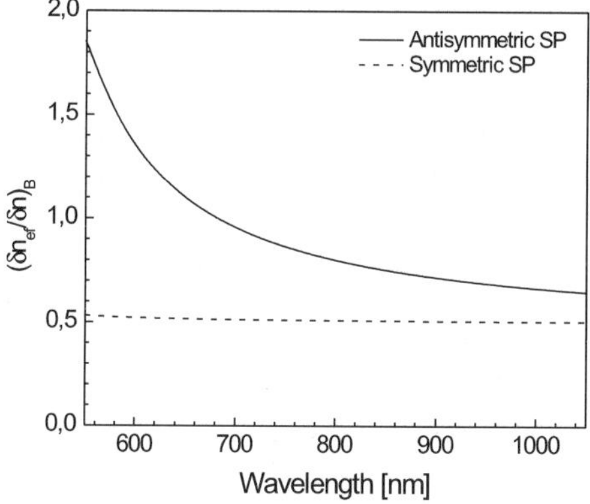

Fig. 18 Sensitivity of the effective index of symmetric and antisymmetric surface plasmons to bulk refractive index changes as a function of wavelength. Waveguide configuration: dielectric ($n_1 = 1.32$)–gold ($2d = 20$ nm)–dielectric superstrate ($n_{\mathrm{d}} = 1.32$)

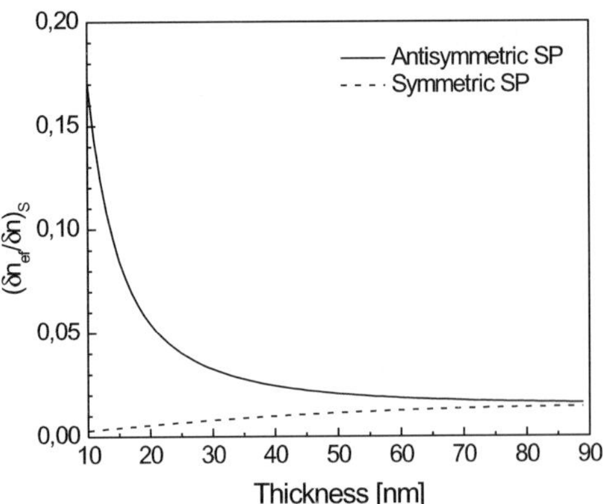

Fig. 19 Sensitivity of the effective index of symmetric and antisymmetric surface plasmons to surface refractive index changes as a function of the thickness of metal film. Waveguide configuration: dielectric ($n_1 = 1.32$)–gold ($\varepsilon_m = -25 + 1.44i$)–thin dielectric film ($h = 5$ nm), dielectric superstrate ($n_d = 1.32$), wavelength 800 nm

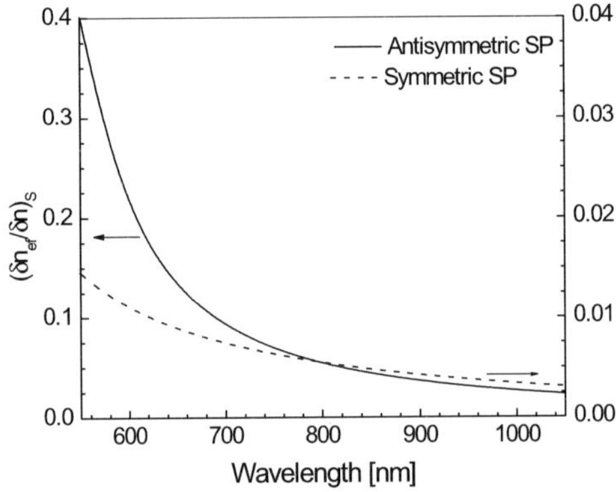

Fig. 20 Sensitivity of the effective index of symmetric and antisymmetric surface plasmons to surface refractive index changes as a function of wavelength. Waveguide configuration: dielectric ($n_1 = 1.32$)–gold ($2d = 20$ nm)–thin dielectric film ($h = 5$ nm), dielectric superstrate ($n_d = 1.32$)

4 Excitation of Surface Plasmons

4.1 Prism Coupling

The most common approach to excitation of surface plasmons is by means of a prism coupler and the attenuated total reflection method (ATR). There are two configurations of the ATR method – Kretschmann geometry [5] and

Otto geometry [4]. In the Kretschmann geometry of the ATR method, a high refractive index prism with refractive index n_p is interfaced with a metal–dielectric waveguide consisting of a thin metal film with permittivity ε_m and thickness q, and a semi-infinite dielectric with a refractive index $n_d (n_d < n_p)$, Fig. 21.

When a light wave propagating in the prism is made incident on the metal film a part of the light is reflected back into the prism and a part propagates in the metal in the form of an inhomogeneous electromagnetic wave [13]. This inhomogeneous wave decays exponentially in the direction perpendicular to the prism–metal interface and is therefore referred as to an evanescent wave. If the metal film is sufficiently thin (less than 100 nm for light in visible and near infrared part of spectrum), the evanescent wave penetrates through the metal film and couples with a surface plasmon at the outer boundary of the metal film. The propagation constant of the surface plasmon propagating along a thin metal film β^{SP} is influenced by the presence of the dielectric on the opposite side of the metal film and can be expressed as

$$\beta^{SP} = \beta^{SP_0} + \Delta\beta = \frac{\omega}{c}\sqrt{\frac{\varepsilon_d \varepsilon_m}{\varepsilon_d + \varepsilon_m}} + \Delta\beta , \tag{65}$$

where β^{SP_0} is the propagation constant of the surface plasmon propagating along the metal–dielectric waveguide in the absence of the prism and $\Delta\beta$ accounts for the finite thickness of the metal film and the presence of the prism. In order for the coupling between the evanescent wave and the surface plasmon to occur, the propagation constant of the evanescent wave β^{EW} and that of the surface plasmon β^{SP} have to be equal:

$$\frac{2\pi}{\lambda} n_p \sin\theta = k_z = \beta^{EW} = \mathrm{Re}\left\{\beta^{SP}\right\} = \mathrm{Re}\left\{\frac{2\pi}{\lambda}\sqrt{\frac{\varepsilon_d \varepsilon_m}{\varepsilon_d + \varepsilon_m}} + \Delta\beta\right\} . \tag{66}$$

In terms of effective index, this coupling condition can be written as follows:

$$n_p \sin\theta = n_{ef}^{EW} = n_{ef}^{SP} = \mathrm{Re}\left\{\sqrt{\frac{\varepsilon_d \varepsilon_m}{\varepsilon_d + \varepsilon_m}}\right\} + \Delta n_{ef}^{SP} , \tag{67}$$

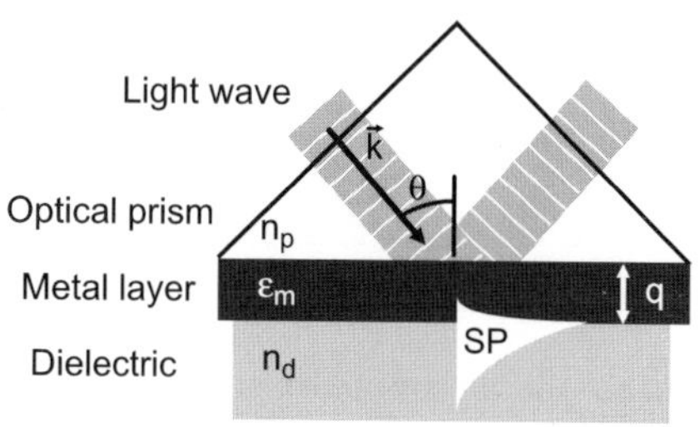

Fig. 21 Excitation of surface plasmons in the Kretschmann geometry of the attenuated total reflection (ATR) method

where n_{ef}^{EW} is the effective index of the evanescent wave, n_{ef}^{SP} is the effective index of the surface plasmon, and $\Delta n_{ef}^{SP} = \mathrm{Re}\{\Delta\beta\lambda/2\pi\}$. The coupling condition between the light wave and the surface plasmon is illustrated in Fig. 22, which shows the spectral dependencies of effective indices of a surface plasmon on a gold–water interface and an evanescent light wave produced by a light wave incident on the gold film from a BK7 glass prism. For each wavelength, the matching condition is satisfied for a single angle of incidence, the coupling angle, which increases with decreasing wavelength.

In the Otto geometry, a high refractive index prism with refractive index n_p is interfaced with a dielectric–metal waveguide consisting of a thin dielectric film with refractive index $n_d (n_d < n_p)$ and thickness q, and a semi-infinite metal with permittivity ε_m, Fig. 23.

In Otto geometry, a light wave incident on the prism–dielectric film interface at an angle of incidence larger than the critical angle of incidence for these two media produces an evanescent wave propagating along the interface between the prism and the dielectric film. If the thickness of the dielectric layer is chosen properly (typically few microns), the evanescent wave and a surface plasmon at the dielectric–metal interface can couple. For the coupling to occur, the propagation constant of the evanescent wave and that of the surface plasmon have to be equal.

The attenuated total reflection method can be also used to excite coupled surface plasmons on thin metal films. The coupling of a light into a symmetric or antisymmetric surface plasmon supported by a thin film (Sect. 2.2) can be in principle achieved in a geometry similar to the Otto geometry (Fig. 23) in which the semi-infinite metal is replaced by a thin metal film [20].

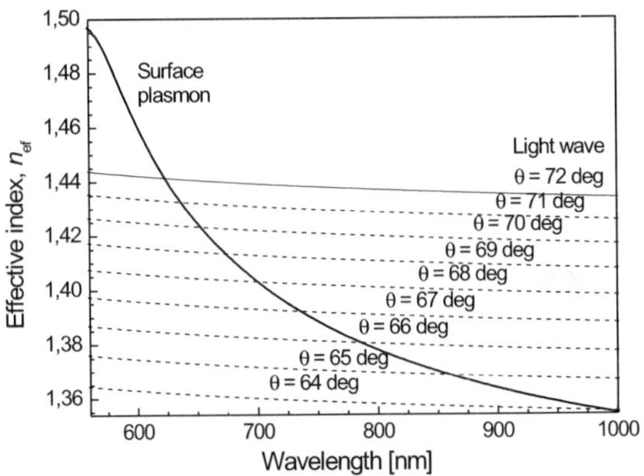

Fig. 22 Spectral dependence of the effective index of a surface plasmon on the interface of gold–water and the effective index of the evanescent light wave produced by a plane light wave incident on the gold film from an optical prism (BK 7 glass) under nine different angles of incidence

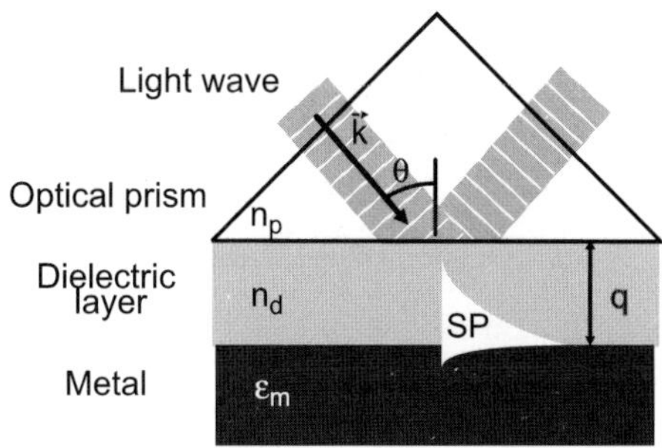

Fig. 23 Excitation of surface plasmons in the Otto geometry of the attenuated total reflection (ATR) method

The interaction between a light wave and a surface plasmon in the ATR method can be investigated using the Fresnel multilayer reflection theory [18]. Herein, we shall present analysis of the reflectivity for the Kretschmann geometry of the ATR method.

Assuming an incident plane wave and a structure prism–metal–dielectric infinite in the *y-z* plane (Fig. 24), the amplitude of reflected light A_{R} can be expressed as:

$$A_{\mathrm{R}} = r_{\mathrm{pmd}} A_{\mathrm{I}} = \left| r_{\mathrm{pmd}} \right| \mathrm{e}^{\mathrm{i}\phi} A_{\mathrm{I}} \,, \tag{68}$$

where A_{I} is the amplitude of the incident light wave, r_{pmd} is an amplitude reflection coefficient and ϕ is a phase shift. The amplitude reflection coefficient is:

$$r_{\mathrm{pmd}} = \frac{r_{\mathrm{pm}} + r_{\mathrm{md}} \exp(2\mathrm{i}k_{\mathrm{mx}} q)}{1 + r_{\mathrm{pm}} r_{\mathrm{md}} \exp(2\mathrm{i}k_{\mathrm{mx}} q)} \,, \tag{69}$$

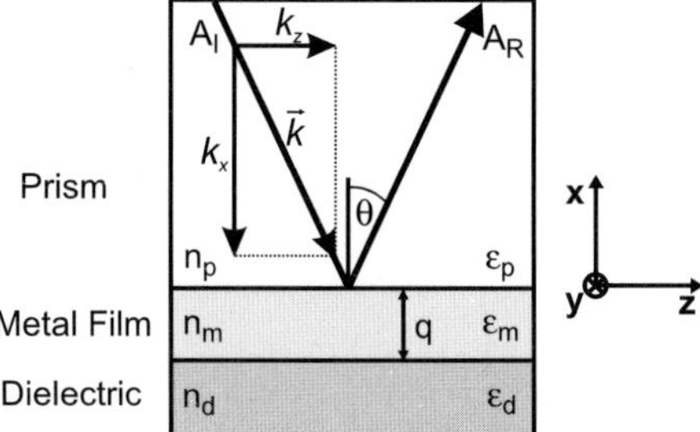

Fig. 24 Light reflection in the Kretschmann geometry of the ATR method

where:

$$k_{ix} = \sqrt{\left(\frac{2\pi}{\lambda}\right)^2 \varepsilon_i - k_z^2}\,, \tag{70}$$

$$r_{ij} = \frac{\varepsilon_j k_{ix} - \varepsilon_i k_{jx}}{\varepsilon_j k_{ix} + \varepsilon_i k_{jx}} \quad \text{for the TM polarization}\,, \tag{71}$$

$$r_{ij} = \frac{k_{ix} - k_{jx}}{k_{ix} + k_{jx}} \quad \text{for the TE polarization}\,, \tag{72}$$

and where subscripts i and j are p, m, or d [19]. Reflectivity (power reflection coefficient) of the structure R is then:

$$R = \left|r_{\text{pmd}}\right|^2 . \tag{73}$$

Figure 25 shows typical dependencies of the reflectivity and phase on the angle of incidence calculated for four different thicknesses of the metal film.

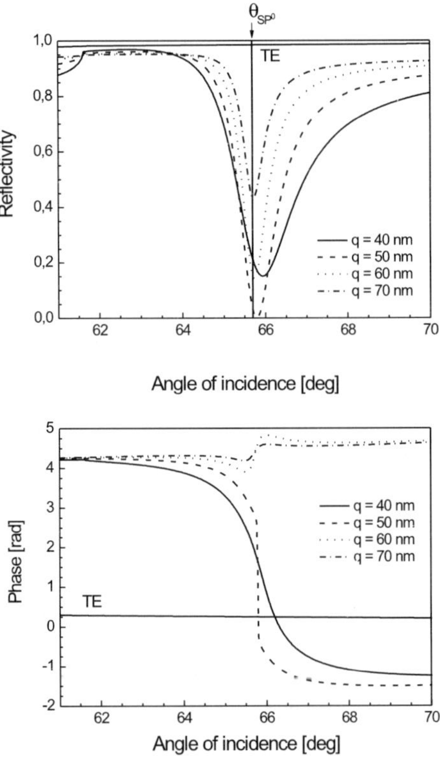

Fig. 25 Reflectivity (*upper plot*) and phase shift (*lower plot*) as a function of the angle of incidence for four different thicknesses of the metal film and TM polarization. Configuration: BK7 glass ($n_\text{p} = 1.51$), gold film ($\varepsilon_\text{m} = -25 + 1.44\text{i}$), water ($n_\text{d} = 1.329$), wavelength 800 nm, reflectivity and phase of the TE polarization are shown for comparison

The angular reflectivity spectra exhibit distinct dips that are associated with the transfer of energy from the incident light wave into a surface plasmon and its subsequent dissipation in the metal film (Fig. 25, upper plot). The interaction between the incident light wave and the surface plasmon also affects the phase of the reflected light, which exhibits an abrupt phase jump [20] (Fig. 25, lower plot). Angular dependencies of the reflectivity and phase of the TE-polarized light contain no resonant features, as no guided modes can be excited by the TE-polarized light in this geometry.

As follows from Fig. 25 (upper plot), the resonant angle of incidence decreases with an increasing metal film thickness and approaches the value θ_{SP_0} corresponding to the coupling of light to a surface plasmon propagating along an isolated metal–dielectric waveguide (Eq. 65, $q \to \infty$, $\Delta\beta = 0$). The depth of the reflectivity dip depends on the thickness of the metal film. The strongest excitation of a surface plasmon ($R = 0$) occurs for a single metal film thickness (for the considered geometry and wavelength, the optimum coupling thickness was about 50 nm). The width and asymmetry of the reflectivity dip increase with a decreasing metal film thickness.

Assuming that the permittivity of metal ε_{m} obeys $|\varepsilon'_{\mathrm{m}}| \gg n_{\mathrm{d}}$ and $|\varepsilon'_{\mathrm{m}}| \gg \varepsilon''_{\mathrm{m}}$, the reflectivity Eq. 69 can be expanded around the resonant value of k_z yielding a Lorentzian (with respect to k_z) approximation of the reflectivity [6]:

$$R(k_z) \doteq 1 - \frac{4\,\mathrm{Im}\left\{\beta^{\mathrm{SP}_0}\right\}\mathrm{Im}\left\{\Delta\beta\right\}}{\left[k_z - \mathrm{Re}\left\{\beta^{\mathrm{SP}}\right\}\right]^2 + (\mathrm{Im}\left\{\beta^{\mathrm{SP}_0}\right\} + \mathrm{Im}\left\{\Delta\beta\right\})^2}, \tag{74}$$

where:

$$\beta^{\mathrm{SP}} = \beta^{\mathrm{SP}_0} + \Delta\beta\,, \tag{75}$$

$$\beta^{\mathrm{SP}_0} = \frac{\omega}{c}\sqrt{\frac{\varepsilon_{\mathrm{d}}\varepsilon_{\mathrm{m}}}{\varepsilon_{\mathrm{d}} + \varepsilon_{\mathrm{m}}}}, \tag{76}$$

$$\Delta\beta = r_{\mathrm{pm}}\,\mathrm{e}^{2ik_{zm}q}2\frac{\omega}{c}\left(\frac{\varepsilon_{\mathrm{d}}\varepsilon_{\mathrm{m}}}{\varepsilon_{\mathrm{d}} + \varepsilon_{\mathrm{m}}}\right)^{3/2}\frac{1}{\varepsilon_{\mathrm{d}} - \varepsilon_{\mathrm{m}}}\,. \tag{77}$$

The term $\Delta\beta$ describes the effect of the prism and, as a complex quantity, has a real part, which perturbs the real part of the propagation constant of a surface plasmon on the interface of semi-infinite dielectric and metal, and an imaginary part, which causes an additional damping of the surface plasmon due to the outcoupling of a portion of the field into the prism [6]. In terms of effective index, the reflectivity (Eq. 74) can be rewritten as follows:

$$R(\theta, \lambda) \doteq 1 - \frac{4\gamma_i\gamma_{\mathrm{rad}}}{(n_{\mathrm{p}}\sin\theta - n_{\mathrm{ef}}^{\mathrm{SP}})^2 + (\gamma_i + \gamma_{\mathrm{rad}})^2}, \tag{78}$$

where $\gamma_i = \mathrm{Im}\left\{\beta^{\mathrm{SP}_0}\right\}\lambda/2\pi$ and $\gamma_{\mathrm{rad}} = \mathrm{Im}\left\{\Delta\beta\right\}\lambda/2\pi$. As follows from Eq. 78, the dip in the reflectivity spectrum is centered at the angle of incidence described by the coupling condition Eq. 67 with $\Delta\beta$ given by Eq. 77. Figure 26 shows the angular reflectivity calculated using the rigorous approach (Eq. 69) and the Lorentzian approximation (Eq. 78) with the propagation constant of a surface plasmon approximated by β^{SP_0} for a model structure: BK7 glass prism, gold film, and water. The approximation provides a good estimate of the position of the reflectivity dip (which would be even closer if the term $\Delta\beta$ was not neglected) and predicts well the shape of the reflectivity curve in the neighborhood of the minimum. In addition, the Lorentzian curve exhibits approximately the same width as the dips calculated using the rigorous approach.

The coupling strength and subsequently the depth of the dip reach the maximum if the radiation and absorption losses of a surface plasmon are equal: $\gamma_i = \gamma_{\mathrm{rad}} = \gamma$. As γ_{rad} decreases with an increasing metal film thickness (as can be deduced from Eq. 77), the condition $\gamma_i = \gamma_{\mathrm{rad}}$ is satisfied only for a single thickness of the metal film, as predicted by the Fresnel reflection theory (Fig. 25). The optimum coupling metal thickness depends on the wavelength and materials involved. For a gold film and wavelengths between 600 and 1000 nm, the optimum coupling thickness varies between 44 nm and 50 nm.

When the optimum coupling occurs ($\gamma_i = \gamma_{\mathrm{rad}} = \gamma$), the angular half-width of the dip $\Delta\theta_{1/2}$ (angular width of the dip at $R = 0.5$) can be expressed from Eq. 78 as:

$$\Delta\theta_{1/2} = \frac{4\gamma}{n_{\mathrm{p}} \cos\theta}\,, \tag{79}$$

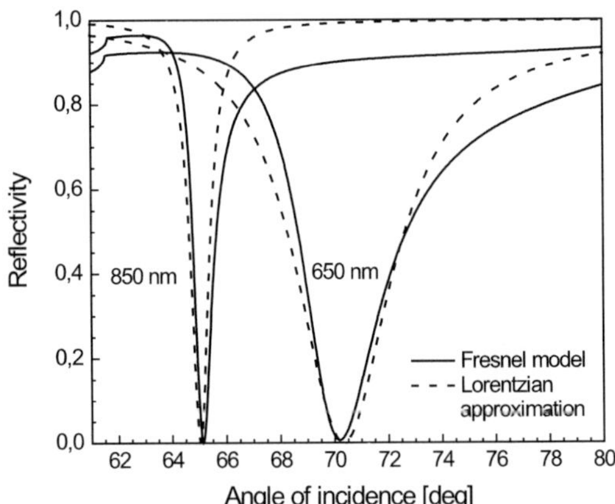

Fig. 26 TM reflectivity as a function of angle of incidence calculated for two different wavelengths using the rigorous Fresnel reflection theory and its Lorentzian approximation. Configuration: BK7 glass, gold film (thickness 48 nm for wavelength 650 nm, and 50 nm for wavelength 850 nm), water

where θ denotes the coupling angle. Equation 79 suggests that the angular width of the dip is proportional to the attenuation of the surface plasmon. As the attenuation coefficient γ decreases rapidly with an increasing wavelength, while the factor $\cos\theta$ changes with the wavelength only slowly, reflectivity dips associated with the excitation of surface plasmons at longer wavelengths (and smaller angles of incidence) are narrower than the dips associated with the excitation of surface plasmons at shorter wavelengths (and higher angles of incidence), Fig. 26.

The characteristic absorption dip can be observed not only in the angular domain, but also when the angle of incidence is kept constant and the wavelength is varied, Fig. 27.

The spectral reflectivity is also described by Eqs. 69 and 74. For low-loss metals ($|\varepsilon'_{\mathrm{m}}| \gg \varepsilon''_{\mathrm{m}}$) with a large real part of the permittivity ($|\varepsilon'_{\mathrm{m}}| \gg \varepsilon_{\mathrm{d}}$), the spectral half-width of the dip $\Delta\lambda_{1/2}$ for the optimum coupling ($\gamma_i = \gamma_{\mathrm{rad}} = \gamma$)

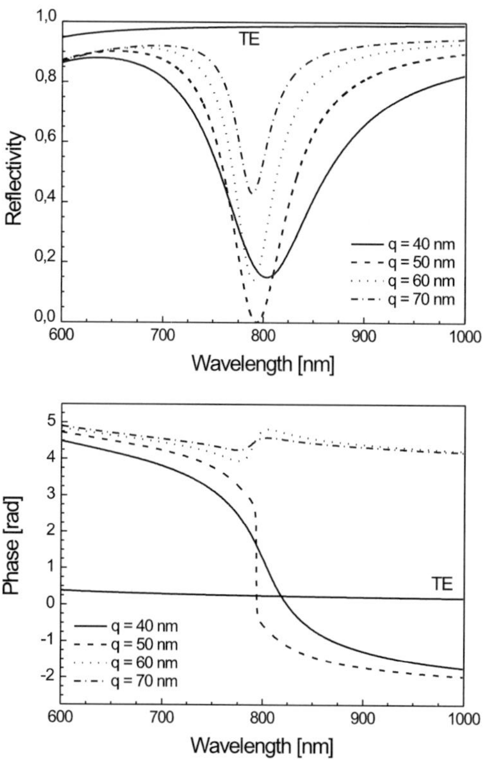

Fig. 27 Reflectivity (*upper plot*) and phase (*lower plot*) of reflected light as a function of wavelength for four different thicknesses of the metal film. Configuration: BK7 glass, gold film, water, angle of incidence 66 deg. Reflectivity and phase for TE polarization are given for comparison

can be calculated from Eq. 74 as:

$$\Delta\lambda_{1/2} = \frac{4\gamma}{\left|\frac{dn_p}{d\lambda}\sin\theta - \frac{dn_{ef}^{SP}}{d\lambda}\right|}, \tag{80}$$

where $dn_p/d\lambda$ is the dispersion of the prism and $dn_{ef}^{SP}/d\lambda$ is the dispersion of the effective index of the surface plasmon. While the attenuation coefficient γ_i decreases with an increasing wavelength, the difference in dispersions

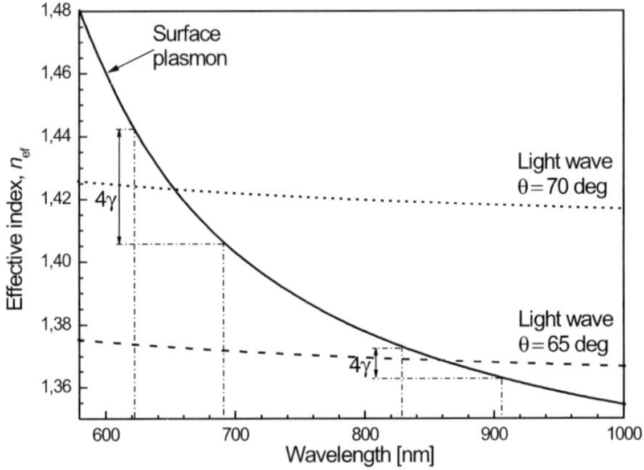

Fig. 28 Spectral dependence of the effective refractive index of a surface plasmon on gold–water interface and the effective index of the evanescent light wave produced in a gold film by a plane light wave incident on the gold film from a BK7 glass prism under two different angles of incidence

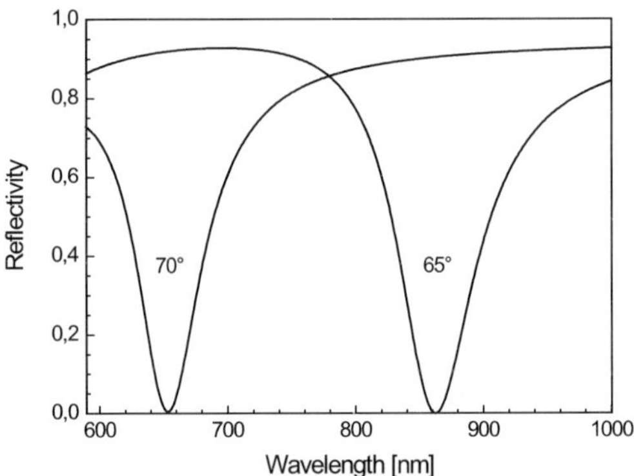

Fig. 29 TM reflectivity as a function of wavelength calculated for two different angles of incidence using the rigorous Fresnel reflection theory. Configuration: BK7 glass, gold film (thickness 48 nm for the wavelength of 650 nm and 50 nm for the wavelength of 850 nm), water

of the effective indices of the evanescent wave and the surface plasmon decreases (Fig. 28) and therefore these two effects can compensate each other. This phenomenon is illustrated in Fig. 29, which shows reflectivity dips produced by the excitation of surface plasmons at the wavelengths of 650 and 850 nm. These dips exhibit approximately the same width although γ_i is about five times larger at the wavelength of 650 nm than at 850 nm.

4.2 Grating Coupling

Another approach to optical excitation of surface plasmons is based on the diffraction of light on a diffraction grating. In this method, a light wave is incident from a dielectric medium with the refractive index n_d on a metal grating with the dielectric constant ε_m, the grating period Λ and the grating depth q, Fig. 30.

When a light wave with the wavevector $\boldsymbol{k}$ is made incident on the surface of the grating, diffraction gives rise to a series of diffracted waves. The wavevector of the diffracted light $\boldsymbol{k}_m$ is:

$$\boldsymbol{k}_m = \boldsymbol{k} + m\boldsymbol{G}\,, \tag{81}$$

where m is an integer and denotes the diffraction order and G is the grating vector [21]. The grating vector lies in the plane of the grating (plane y-z in Fig. 30) and is perpendicular to the grooves of the grating. Its magnitude is inversely proportional to the pitch of the grating and therefore, for the grating geometry considered herein, it can be expressed as:

$$\boldsymbol{G} = \frac{2\pi}{\Lambda}\boldsymbol{z}_0\,. \tag{82}$$

Therefore the component of the wavevector of the diffracted light perpendicular to the plane of the grating k_{xm} is equal to that of the incident wave while the component of the wavevector in the plane of the grating k_{zm} is diffraction altered:

$$k_{zm} = k_z + m\frac{2\pi}{\Lambda}\,. \tag{83}$$

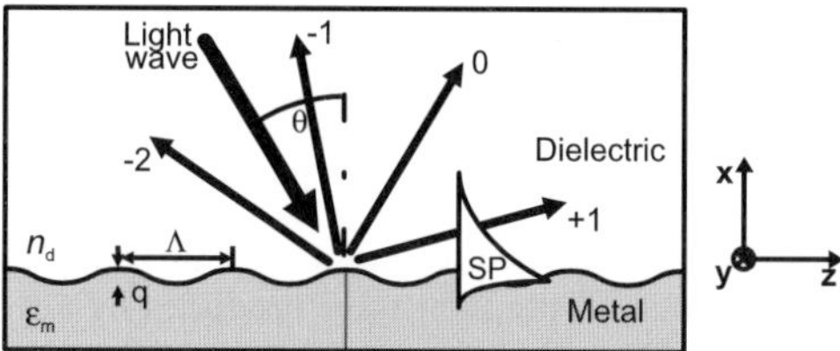

Fig. 30 Excitation of surface plasmons by the diffraction of light on a diffraction grating

The diffracted waves can couple with a surface plasmon when the propagation constant of the diffracted wave propagating along the grating surface k_{zm} and that of the surface plasmon β^{SP} are equal:

$$\frac{2\pi}{\lambda} n_{\mathrm{d}} \sin\theta + m\frac{2\pi}{\Lambda} = k_{zm} = \pm \operatorname{Re}\left\{\beta^{\mathrm{SP}}\right\} , \tag{84}$$

where:

$$\beta^{\mathrm{SP}} = \beta^{\mathrm{SP}_0} + \Delta\beta = \frac{\omega}{c}\sqrt{\frac{\varepsilon_{\mathrm{d}}\varepsilon_{\mathrm{m}}}{\varepsilon_{\mathrm{d}}+\varepsilon_{\mathrm{m}}}} + \Delta\beta , \tag{85}$$

and β^{SP_0} denotes the propagation constant of the surface plasmon propagating along the smooth interface of a semi-infinite metal and a semi-infinite dielectric, and $\Delta\beta$ accounts for the presence of the grating. In terms of effective index, the coupling condition can be rewritten as:

$$n_{\mathrm{d}} \sin\theta + m\frac{\lambda}{\Lambda} = \pm\left(\operatorname{Re}\left\{\sqrt{\frac{\varepsilon_{\mathrm{d}}\varepsilon_{\mathrm{m}}}{\varepsilon_{\mathrm{d}}+\varepsilon_{\mathrm{m}}}}\right\} + \Delta n_{\mathrm{ef}}^{\mathrm{SP}}\right) , \tag{86}$$

where $\Delta n_{\mathrm{ef}}^{\mathrm{SP}} = \operatorname{Re}\{\Delta\beta\lambda/2\pi\}$.

The coupling condition between a diffracted light wave and a surface plasmon is illustrated in Fig. 31. The effective index of light diffracted on two different gratings ($\Lambda = 540$ nm and $\Lambda = 672$ nm) is diffraction enhanced to match the effective index of a surface plasmon on a gold–water interface. As illustrated in Fig. 31, different orders of diffraction (first order for the grating with $\Lambda = 672$ nm and minus first order for the grating with $\Lambda = 540$ nm) can be used to fulfill the matching condition. The effective index of the inhomoge-

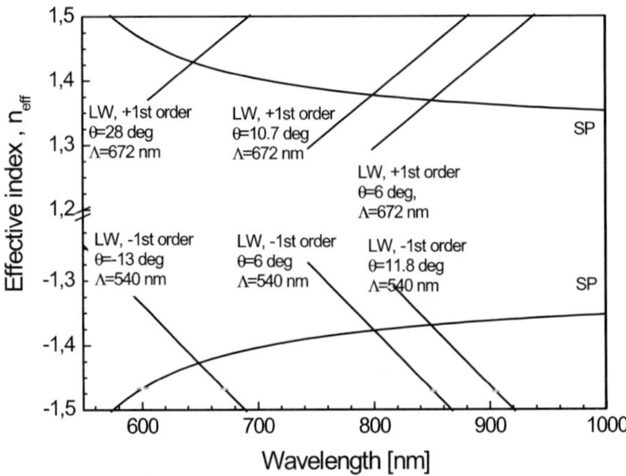

Fig. 31 Spectral dependence of the effective index of a surface plasmon on gold–water interface and the effective index of the light wave produced by a diffraction of light on a diffraction grating calculated for two different grating periods and three different angles of incidence

neous light wave is approximately a linear function of wavelength with a slope equal to m/Λ, which is positive for $m > 0$ and negative for $m < 0$. The coupling condition Eq. 86 can be fulfilled for various combinations of the angle of incidence, grating pitch, and diffraction order. For the positive diffraction orders, the coupling wavelength increases with a decreasing angle of incidence, while for the negative diffraction orders, the coupling wavelength increases with an increasing angle of incidence.

The grating-moderated interaction between a light wave and a surface plasmon can be investigated by solving Maxwell's equations in differential or integral form. In the differential method, the grating profile is approximated with a stack of layers in which a solution of the Maxwell equations is calculated in the form of a Rayleigh series and the total solution of the diffraction problem is found by applying boundary conditions at each interface [22, 23]. The integral method assumes a certain current flow at the grating surface

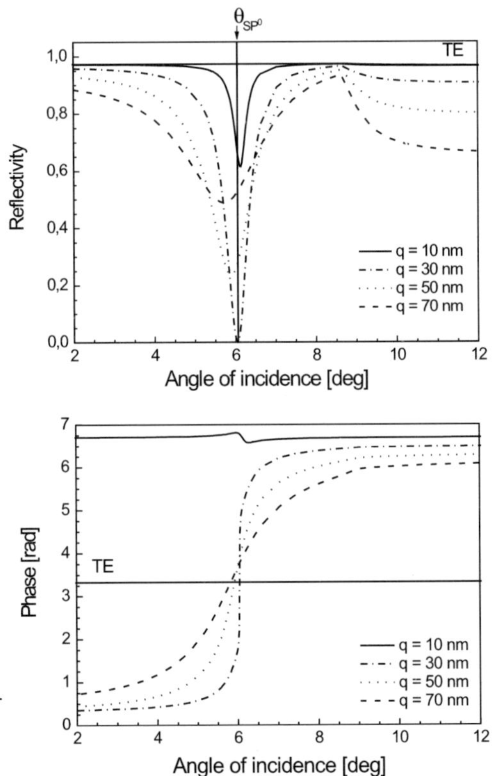

Fig. 32 Reflectivity (*upper plot*) and phase (*lower plot*) as a function of the angle of incidence for four different depths of a metallic sinusoidal grating and TM polarization. Configuration: gold ($\varepsilon_m = -25 + 1.44i$), water ($n_d = 1.329$), wavelength 800 nm, grating period 540 nm, angle of incidence taken in air. Reflectivity and phase of the TE polarization are shown for comparison

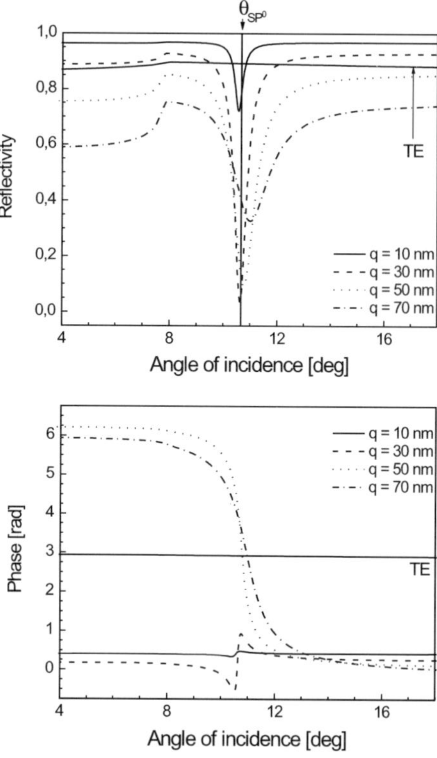

Fig. 33 Reflectivity (*upper plot*) and phase (*lower plot*) as a function of the angle of incidence for four different depths of metallic sinusoidal grating and TM polarization. Configuration: gold ($\varepsilon_m = -25 + 1.44i$), water ($n_d = 1.329$), wavelength 800 nm, grating period 672 nm, angle of incidence taken in air. Reflectivity and phase of the TE polarization are shown for comparison

and reduces the problem to the calculation of the Helmholtz–Kirchhoff integral [24]. Figures 32 and 33 show the dependence of the reflectivity and phase on the angle of incidence for light incident from water onto a gold grating and two different grating pitches, $\Lambda = 540$ nm (Fig. 32) and $\Lambda = 672$ nm (Fig. 33), and four different grating depths. These spectra were calculated using the integral method. The angular reflectivity spectra (upper plots in Fig. 32 and Fig. 33) exhibits a characteristic dip caused by the transfer of energy of the incident light into a surface plasmon. On shallow diffraction gratings, surface plasmons are excited at the angles of incidence close to the coupling angles predicted from the matching condition, neglecting the effect of the grating ($q \to 0$ and $\Delta n_{ef}^{SP} = \mathrm{Re}\{\Delta\beta\} = 0$), Fig. 31. The coupling angle of incidence decreases with an increasing depth of the grating when the surface plasmons are excited by a negative order of diffraction, and follows an opposite trend when the surface plasmons are excited by a positive order of diffraction. The depth of the reflectivity dip depends on the depth of the grating and the strongest excitation of a surface plasmon ($R = 0$) occurs for a single depth of the grating (for the considered geometry and wavelength, the optimum grating depth is about 30 nm). The width and asymmetry of the reflectivity dip increase with an increasing depth of grating. The interaction between the light wave and the surface plasmon results also in a change in the phase of the reflected light, Fig. 32 and Fig. 33 (lower plot).

The characteristic absorption dip can be observed not only in the angular domain, but also when the angle of incidence is kept constant and the wavelength is varied, as illustrated in Fig. 34 and Fig. 35.

Figure 36 shows the angular reflectivity for light incident from water onto a gold grating. The dips produced at the wavelength of 850 nm are about five times narrower than those occurring at 650 nm. The ratio of the dip widths corresponds to the ratio of the attenuation coefficients for surface plasmons

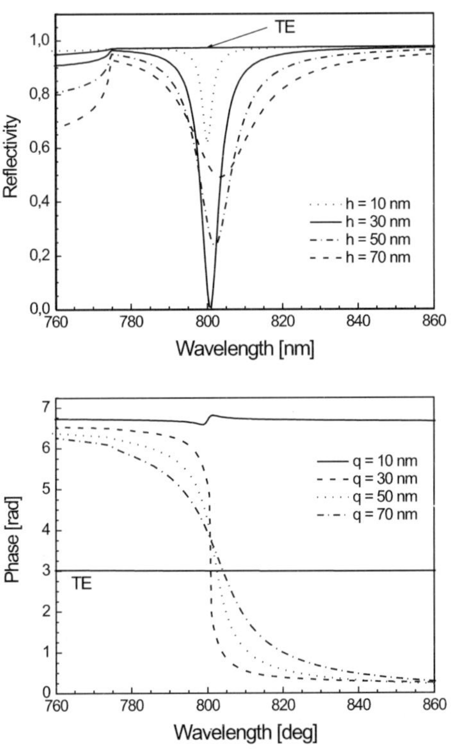

Fig. 34 Reflectivity (*upper plot*) and phase (*lower plot*) as a function of the wavelength for four different modulation depths of metallic sinusoidal grating and TM polarization. Configuration: gold–water, angle of incidence 6 degrees, grating period 540 nm, angle of incidence taken in air. Reflectivity and phase of the TE polarization are shown for comparison

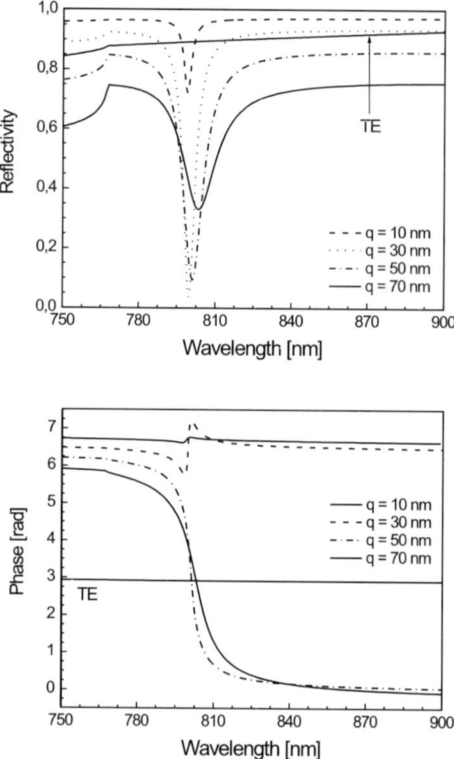

Fig. 35 Reflectivity (*upper plot*) and phase (*lower plot*) as a function of the wavelength for four different modulation depths of metallic sinusoidal grating and TM polarization. Configuration: gold–water, angle of incidence 10.7 degrees, grating period 672 nm, angle of incidence taken in air. Reflectivity and phase of the TE polarization are shown for comparison

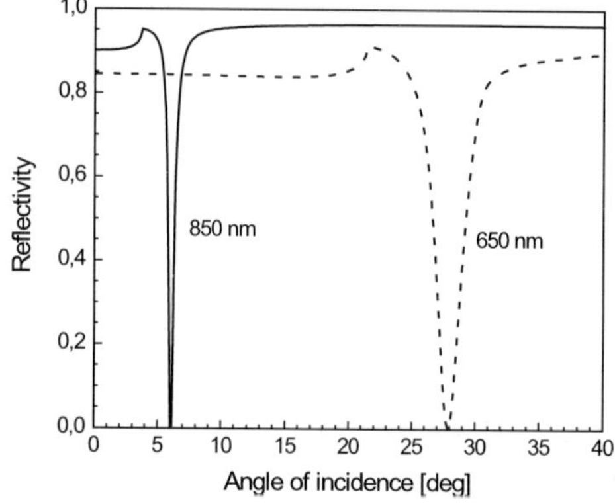

Fig. 36 Reflectivity as a function of angle of incidence calculated for two wavelengths. Configuration: gold–water interface, grating period 672 nm, angle of incidence taken in air

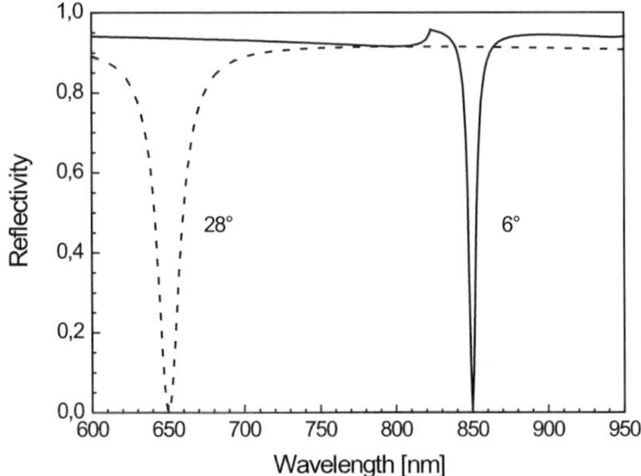

Fig. 37 Reflectivity as a function of wavelength calculated for two different angles of incidence. Configuration: gold–water interface, grating period 672 nm, angle of incidence taken in air

at 650 and 850 nm, as in the case of prism coupling (Sect. 4.1). However, the width of the dips observed in the wavelength spectrum (Fig. 37) varies with the wavelength, which contrasts with the weak dependence of the width of spectral dips in the case of prism coupling. This effect can be attributed to the fact that the difference in the dispersions of the effective indices of the evanescent wave and surface plasmon is large (Fig. 31) and varies relatively little over the considered wavelength range.

4.3 Waveguide Coupling

Surface plasmons can be also excited by modes of a dielectric waveguide. An example of a waveguiding structure integrating a dielectric waveguide and a metal–dielectric waveguide is shown in Fig. 38. A mode of the dielectric waveguide propagates along the waveguide and when it enters the region with

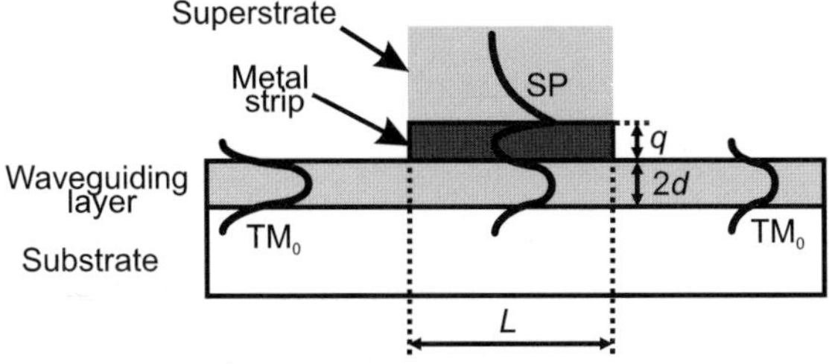

Fig. 38 Excitation of surface plasmons by a mode of a dielectric waveguide

a thin metal film, it penetrate through the metal film and couples with a surface plasmon at the outer boundary of the metal.

The coupling between the waveguide mode and a surface plasmon can occur when the propagation constant of the mode β_M is equal to the real part of the propagation constant of the surface plasmon β_{SP}:

$$\beta_M = \mathrm{Re}\,\{\beta_{SP}\}\ . \tag{87}$$

The coupling between the waveguide mode and a surface plasmon can be investigated by analyzing hybrid modes, which are solutions of the vector wave (Eq. 18) for the coupled waveguides [25]. The propagation of light through the entire waveguiding structure can be simulated using the mode expansion and propagation method [26]. In this method, the simulated waveguide is subdivided into longitudinally uniform sections and, in each section, a set of eigenmodes is calculated. The mutual relationships among modal amplitudes at both sides of the interface between the longitudinal sections are obtained from the continuity of the transversal field components by mode matching [26].

As surface plasmons are typically much more dispersive than modes of common dielectric waveguides, the coupling condition Eq. 87 is fulfilled only for a narrow range of wavelengths. Therefore, the excitation of a surface plasmon can be observed as a narrow dip in the spectrum of transmitted light, Fig. 39. The strength of the coupling depends on the metal thickness (Fig. 39) and the length of the interaction region (Fig. 40). The effect of the metal film thickness and interaction length is depicted in Figs. 39 and 40 for a model structure consisting of substrate (refractive index 1.514), a waveguiding layer (refractive index 1.517, thickness 3 μm), a thin gold layer, and superstrate (refractive index 1.40).

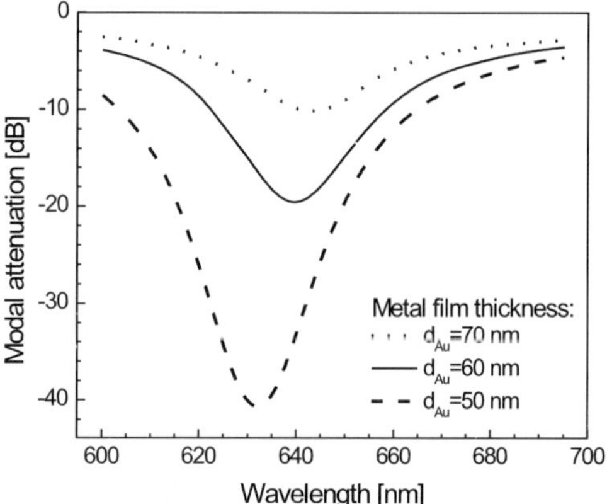

Fig. 39 Spectral dependence of the transmission of a slab waveguide with a thin metal strip for different thicknesses of the metal film, metal strip length $L = 1$ mm

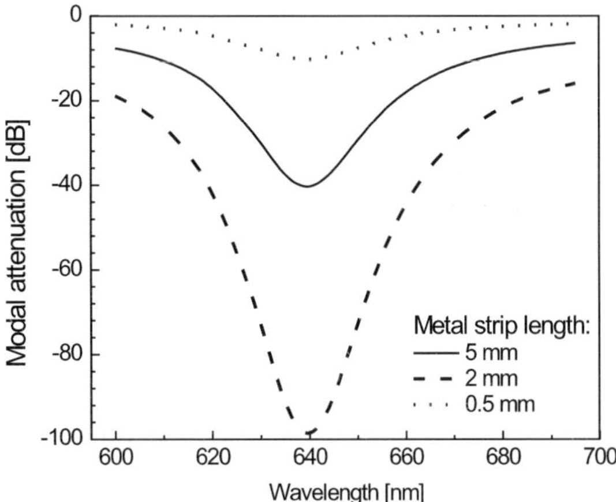

Fig. 40 Spectral dependence of the transmission of a slab waveguide with a thin metal strip for different lengths of the metal strip, metal film thickness $q = 60$ nm

5 Summary

Surface plasmons are special modes of electromagnetic field in metal–dielectric waveguides. They are characterized by the field distribution and complex propagation constant, which can be determined from an appropriate eigenvalue equation. The propagation constant of surface plasmons is highly sensitive to changes in the refractive index distribution, as can be demonstrated using the perturbation theory. Surface plasmons can be excited by light waves using (i) prism coupling and the attenuated total reflection, (ii) diffraction on a metal diffraction grating, and (iii) coupling among parallel optical waveguides.

References

1. Wood RW (1902) Philosophical Magazine 4:396
2. Fano U (1941) J Opt Soc Am 31:231
3. Turbadar T (1959) Proc Phys Soc 73:40
4. Otto A (1968) Zeits Phys 216:398
5. Kretschmann E, Raether H (1968) Z Naturforsch 2135–2136:2135–2136
6. Raether H (1988) Springer Tracts Mod Phys 111:1
7. Boardman AD (1982) Electromagnetic surface modes. Wiley, Chichester
8. Pockrand I, Swalen JD, Gordon JG, Philpott MR (1978) Surf Sci 74:237
9. Gordon JG, Ernst S (1980) Surf Sci 101:499
10. Snyder AW, Love JD (1983) Optical waveguide theorie, Chap viii. Science paperbacks 190. Chapman and Hall, London New York, p 734
11. Marcuse D (1973) Integrated optics. IEEE Press, New York

12. Marcuse D (1974) Theory of dielectric optical waveguides. Academic, New York
13. Born M, Wolf E (1999) Principles of optics: electromagnetic theory of propagation, interference and diffraction of light. Cambridge University Press, Cambridge
14. Palik ED, Ghosh G (1998) Handbook of optical constants of solids. Academic, San Diego
15. Stegeman GI, Burke JJ, Hall DG (1983) Optics Lett 8:383
16. Burke JJ, Stegeman GI, Tamir T (1986) Phys Rev B 33:5186
17. Sarid D (1981) Phys Rev Lett 47:1927
18. Tobiška P (2005) PhD Dissertation, Charles University, Prague
19. Yariv A, Yeh P (2003) Optical waves in crystals: propagation and control of laser radiation. Wiley, Hoboken NJ
20. Deck RT, Sarid D, Olson GA, Elson JM (1983) Appl Opt 22:3397
21. Grigorenko AN, Nikitin PI, Kabashin AV (1999) Appl Phys Lett 75:3917
22. Hutley MC (1982) Diffraction gratings. Academic, London
23. Moharam MG, Gaylord TK (1986) J Opt Soc Am A 3:1780
24. Li LF (1993) J Opt Soc Am A 10:2581
25. Goray LI, Seely JF (2002) Appl Opt 41:1434
26. Homola J (1997) Sens Act B Chem 39:286
27. Čtyroký J, Homola J, Skalský M (1997) Opt Quant Electron 29:301

Springer Ser Chem Sens Biosens (2006) 4: 45–67
DOI 10.1007/5346_014

Published online: 8 July 2006

Surface Plasmon Resonance (SPR) Sensors

Jiří Homola (✉) · Marek Piliarik

Institute of Radio Engineering and Electronics, Prague, Czech Republic
homola@ure.cas.cz

Keywords Affinity biosensor · Guided mode · Optical sensor · Spectroscopy of guided modes · Surface plasmon resonance sensor

1 Introduction

In Chap. 1 by J. Homola of this volume [1] surface plasmons were introduced as modes of dielectric/metal planar waveguides and their properties were established. It was demonstrated that the propagation constant of a surface plasmon is sensitive to variations in the refractive index at the surface of a metal film supporting the surface plasmon. In this chapter, it is shown how this phenomenon can be used to create a sensing device. The concept of optical sensors based on surface plasmons, commonly referred as to surface plasmon resonance (SPR) sensors, is described and the main approaches to SPR sensing are presented. In addition, the concept of affinity biosensors is introduced and the main performance characteristics of SPR biosensors are defined.

2
Surface Plasmon Resonance (SPR) Sensors

An optical sensor is a sensing device which, by optical means, converts the quantity being measured (measurand) to another quantity (output) which is typically encoded into one of the characteristics of a light wave. In SPR sensors, a surface plasmon is excited at the interface between a metal film and a dielectric medium (superstrate), changes in the refractive index of which are to be measured. A change in the refractive index of the superstrate produces a change in the propagation constant of the surface plasmon. This change alters the coupling condition between a light wave and the surface plasmon, which can be observed as a change in one of the characteristics of the optical wave interacting with the surface plasmon [2]. Based on which characteristic of the light wave interacting with the surface plasmon is measured, SPR sensors can be classified as SPR sensors with angular, wavelength, intensity, phase, or polarization modulation (Fig. 1).

In *SPR sensors with angular modulation* a monochromatic light wave excites a surface plasmon. The strength of coupling between the incident wave and the surface plasmon is observed at multiple angles of incidence of the light wave and the angle of incidence yielding the strongest coupling is measured and used as a sensor output (Fig. 2, upper plot) [3]. The sensor output can be calibrated to refractive index. In *SPR sensors with wavelength modulation*, a surface plasmon is excited by a collimated light wave containing multiple wavelengths. The angle at which the light wave is incident onto the metal film is kept constant. The strength of coupling between the incident wave and the surface plasmon is observed at multiple wavelengths and the wavelength yielding the strongest coupling is measured and used as a sensor output (Fig. 2, lower plot) [4]. *SPR sensors with intensity modulation* are

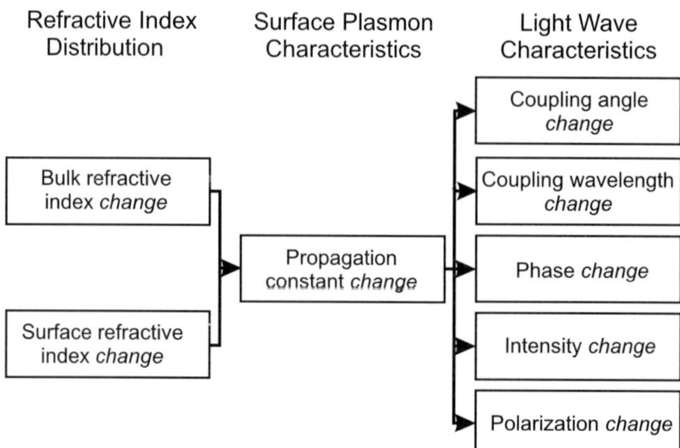

Fig. 1 Concept of surface plasmon resonance sensors

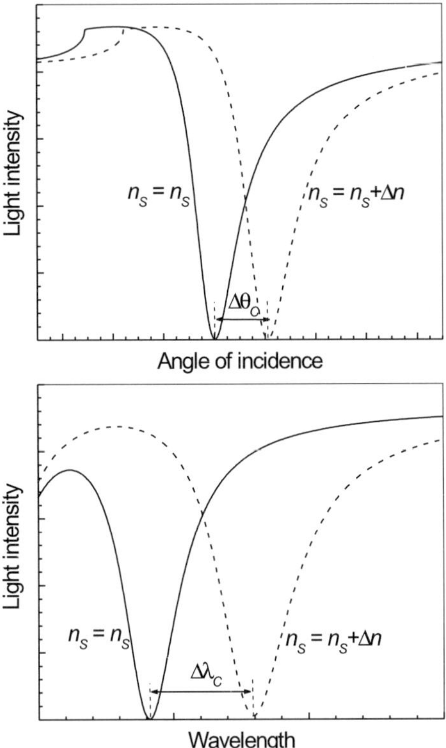

Fig. 2 Intensity of light wave interacting with a surface plasmon as a function of angle of incidence (*upper plot*) and wavelength (*lower plot*) for two different refractive indices of superstrate

based on measuring the strength of the coupling between the incident light wave and a surface plasmon at a single angle of incidence and wavelength and the intensity of light wave serves as a sensor output [5]. In *SPR sensors with phase modulation* the shift in phase of the light wave interacting with the SP is measured at a single angle of incidence and wavelength of the light wave and used as a sensor output [6]. In *SPR sensors with polarization modulation*, changes in the polarization of the light wave interacting with a surface plasmon are measured [7].

SPR sensors are either direct or indirect. In direct SPR sensors, the measurand (typically refractive index) modulates characteristics of the light directly. In indirect SPR sensors, the measurand modulates an intermediate quantity which then modulates the light characteristics. SPR affinity biosensors are a typical example of indirect SPR sensors.

3
Surface Plasmon Resonance Affinity Biosensors

SPR affinity biosensors are SPR sensing devices incorporating biorecognition elements (e.g., antibodies) that recognize and are able to interact with a selected analyte. The biorecognition elements are immobilized on the surface of a metal film supporting a surface plasmon. When a solution containing analyte molecules is brought into contact with the SPR sensor, analyte molecules in solution bind to the molecular recognition elements, producing an increase in the refractive index at the sensor surface. This change gives rise to a change in the propagation constant of the surface plasmon (Fig. 3). The change in the propagation constant is determined by measuring a change in one of the characteristics of the light wave interacting with the surface plasmon (Fig. 1) [2].

The amount of the refractive index change Δn_b induced by the analyte molecules binding to the biorecognition elements depends on the volume refractive index increment $(\mathrm{d}n/\mathrm{d}c)_{vol}$ and can be expressed as:

$$\Delta n_b = \left(\frac{\mathrm{d}n}{\mathrm{d}c}\right)_{vol} \Delta c_b , \tag{1}$$

where Δc_b is the concentration of bound analyte expressed in mass/volume. The value of the refractive index increment depends on the structure of the analyte molecules and varies from 0.1 to 0.3 $\mathrm{mL\,g^{-1}}$ [8, 9]. If the binding occurs within a thin layer at the sensor surface of thickness h the refractive index change can be rewritten as:

$$\Delta n_b = \left(\frac{\mathrm{d}n}{\mathrm{d}c}\right)_{vol} \frac{\Delta \Gamma}{h} , \tag{2}$$

where Γ denotes the surface concentration in mass/area [10].

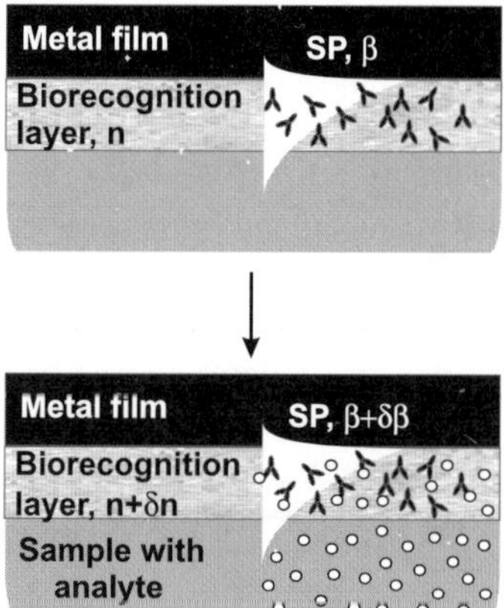

Fig. 3 Principle of surface plasmon resonance affinity biosensor

4 Main Performance Characteristics of SPR Sensors

In direct (refractometric) SPR sensors, refractive index (measurand) directly modulates characteristics of the light wave such as the coupling angle or wavelength, intensity, phase, and polarization (sensor output) (Fig. 4, upper diagram). In SPR affinity biosensors, the measurand is usually a concentration of a chemical or biological analyte, which through the binding of analyte to a biorecognition element is converted into a refractive index change at the sensor surface, which then modulates characteristics of the light wave (sensor output) (Fig. 4, lower diagram).

The sensor response to a given value of the measurand can be predicted by the sensor transfer function F, $Y = F(X)$ determined from a theoretical sensor model or a sensor calibration. However, the value of the measurand determined by the sensor X_{measured} differs from the true value of the measurand X_{true}:

$$X_{\text{measured}} = X_{\text{true}} + e\,, \tag{3}$$

by the measurement error e. There are various sources of error. Random errors are statistical fluctuations (in either direction) in the measured data due to the precision limitations of the sensor system. Random errors are not eliminated by calibration. Systematic errors, by contrast, are reproducible inaccuracies that are consistently in the same direction. Systematic errors are reduced by calibration to the uncertainty level of the calibration system. The uncertainty of a calibration depends primarily on the accuracy of the reference(s) and stability of the test environment.

The main characteristics describing the performance of SPR (bio)sensors include sensitivity, linearity, resolution, accuracy, reproducibility, dynamic range, limit of detection, and limit of quantification.

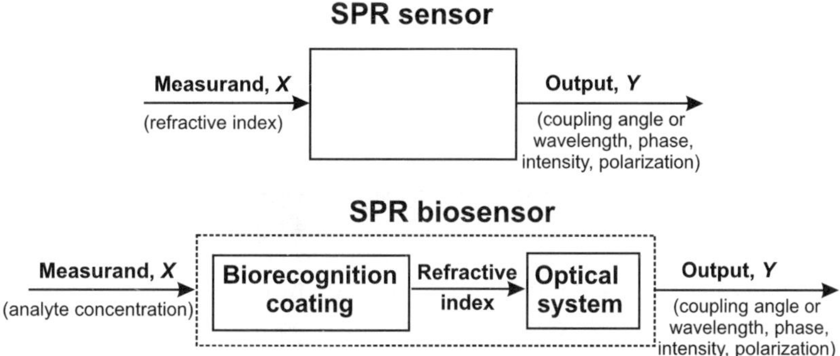

Fig. 4 Direct and indirect SPR sensors: measurand and sensor output

Sensor *sensitivity* is the ratio of the change in sensor output to the change in the measurand (slope of the calibration curve):

$$S = \frac{\partial Y}{\partial X} . \tag{4}$$

Refractometric sensitivity describes the sensitivity of the SPR sensor to refractive index n and can be written as:

$$S_{\text{RI}} = \frac{\partial Y}{\partial n} . \tag{5}$$

Sensitivity of an SPR biosensor to a concentration of analyte c can be written as:

$$S_{\text{c}} = \frac{\partial Y}{\partial c} . \tag{6}$$

Sensitivity of SPR sensors will be discussed in detail in Sect. 4.1.

Sensor *linearity* may concern primary measurand (concentration of analyte) or refractive index and defines the extent to which the relationship between the measurand and sensor output is linear over the working range. Linearity is usually specified in terms of the maximum deviation from a linear transfer function over the specified dynamic range. In general, sensors with linear transfer functions are desirable as they require fewer calibration points to produce an accurate sensor calibration. However, response of SPR biosensors is usually a non-linear function of the analyte concentration and therefore calibration needs to be carefully considered.

The *resolution* of a sensor is the smallest change in measurand which produces a detectable change in the sensor output. In SPR sensors, the term resolution usually refers to a bulk refractive index resolution. In SPR biosensors, an equivalent of this term is the limit of detection described below. Resolution of SPR sensors will be discussed in detail in Sect. 4.2.

Sensor *accuracy* describes the closeness of agreement between a measured value and a true value of the measurand (concentration of analyte or refractive index). Sensor accuracy is usually expressed in absolute terms or as a percentage of the error/output ratio.

Reproducibility is the ability of the sensor to provide the same output when measuring the same value of the measurand (concentration of analyte or refractive index) under the same operating conditions over a period of time. The reproducibility is typically expressed as the percentage of full range.

The (*dynamic*) *range* describes the span of the values of the measurand that can be measured by the sensor. In refractometric SPR sensors the dynamic range usually describes a range of values of the refractive index of the sample that can be measured with a specified accuracy. Dynamic range of SPR biosensors defines the range of concentrations of an analyte which can be measured with specified accuracy and extends from the lowest concentration at which a quantitative measurement can be done, i.e., the limit of quantification.

An important characteristic describing the ability of a biosensor to detect an analyte is the *limit of detection* (LOD). LOD, as defined by the International Union of Pure and Applied Chemistry, is the concentration of analyte c_L derived from the smallest measure Y_L that can be detected with reasonable certainty. The value of Y_LOD is given by the equation:

$$Y_\mathrm{LOD} = Y_\mathrm{blank} + m\sigma_\mathrm{blank} \,, \tag{7}$$

where Y_blank is the mean of the blank (sample with no analyte) measures, σ_blank is the standard deviation of the blank measures, and m is a numerical factor chosen according to the confidence level desired (typically 2 or 3) [11]. As $c_\mathrm{blank} = 0$, the LOD concentration c_LOD can be expressed as:

$$c_\mathrm{LOD} = \frac{1}{S_\mathrm{c}(c=0)} m\sigma_\mathrm{blank} \,, \tag{8}$$

where S_c denotes the sensor sensitivity to analyte concentration. It should be noted that this definition of LOD inherently recognizes only the false positives, which in effect makes the probability of a false negative equal to 50% [11]. Therefore, another approach to the definition of LOD can be found in the literature, employing two independent values (each equal to 0.05 or 0.01) for the probability of the false positives and negatives [12].

As the LOD defined by Eq. 8 defines the concentration at which one can decide whether the analyte is present, rather than quantifying the analyte concentration, another performance characteristic – the *limit of quantification* (LOQ) – is sometimes used [11]. Analyte quantification is generally accepted to begin at a concentration equal to 10 standard deviations of the blank. Thus, the LOQ concentration c_LOQ, can be expressed as:

$$c_\mathrm{LOQ} = \frac{10}{S_\mathrm{c}(c=0)} \sigma_\mathrm{blank} \,. \tag{9}$$

4.1 Sensitivity

The sensitivity of an SPR biosensor can be decomposed into two contributions:

$$S_\mathrm{c} = \frac{\partial Y}{\partial c} = \frac{\partial Y}{\partial n_\mathrm{b}} \frac{\mathrm{d}n_\mathrm{b}(c)}{\mathrm{d}c} = S_\mathrm{RI} S_\mathrm{nc} \,, \tag{10}$$

where S_RI denotes the sensitivity of the output to a refractive index profile change and S_nc is derived from the refractive index change (n_b) caused by the binding of analyte (concentration c) to biorecognition elements. The sensitivity of an SPR sensor to a refractive index S_RI can be also broken down into two contributions:

$$S_{\mathrm{RI}} = \frac{\delta Y}{\delta n_{\mathrm{ef}}} \frac{\delta n_{\mathrm{ef}}}{\delta n_{\mathrm{b}}} = S_{\mathrm{RI1}} S_{\mathrm{RI2}} \,. \tag{11}$$

The first term S_{RI1} depends on the method of excitation of surface plasmons and the modulation approach used in the SPR sensor and is hereafter referred as to the instrumental contribution. S_{RI2} describes the sensitivity of the effective index of a surface plasmon to refractive index and is independent of the modulation method and the method of excitation of surface plasmons. The sensitivity of surface plasmon to refractive index S_{RI2} depends on the profile of the refractive index n_{b} and has been analyzed in Chap. 1 of this volume [1] for the two main types of refractive index changes – surface refractive index change and bulk refractive index change.

In the following section, sensitivity of the sensor output to effective index of a surface plasmon is analyzed for selected sensor configurations, and the merit of different SPR sensor configurations in terms of bulk refractive index sensitivity is evaluated.

4.1.1 Sensitivity of SPR Sensors with Angular Modulation

In SPR sensors with angular modulation, the sensor output is the coupling angle θ_{r} and therefore the instrumental contribution to sensor sensitivity S_{RI1} is equal to $\delta\theta_{\mathrm{r}}/\delta n_{\mathrm{ef}}$. After a straightforward manipulation, differentiation of the coupling conditions for the prism coupler (see Chap. 1 of this volume [1]) in n_{ef} and θ (θ_i for the grating coupler, the angle of incidence is given in a medium with a refractive index $n_i = 1$), yields:

$$\left(\frac{\delta\theta_{\mathrm{r}}}{\delta n_{\mathrm{ef}}}\right)_{\mathrm{prism}} = \frac{1}{n_{\mathrm{p}} \cos\theta_{\mathrm{r}}} = \frac{1}{\sqrt{n_{\mathrm{p}}^2 - n_{\mathrm{ef}}^2}} \,, \tag{12}$$

where n_{p} is the refractive index of the prism. Similarly, by differentiating the coupling condition for the grating coupler (see Chap. 1 in this volume [1]), one can obtain:

$$\left(\frac{\delta\theta_{\mathrm{r}}}{\delta n_{\mathrm{ef}}}\right)_{\mathrm{grating}} = \frac{\mathrm{sgn}(m)}{\cos\theta_{\mathrm{r}}} = \frac{\mathrm{sgn}(m)}{\sqrt{1 - \left(n_{\mathrm{ef}} - \frac{|m|\lambda}{\Lambda}\right)^2}} \,, \tag{13}$$

where m denotes the order of diffraction. Equation 13 suggests that $(\delta\theta_{\mathrm{r}}/\delta n_{\mathrm{ef}})_{\mathrm{grating}}$ is positive for positive diffraction orders and negative for negative diffraction orders (resonant angle decreases with n_{ef} increase). An increase in the diffraction order can be compensated for by a decrease in the grating period Λ. Figure 5 depicts the wavelength dependence of $\delta\theta_{\mathrm{r}}/\delta n_{\mathrm{ef}}$ for model prism- and grating-based SPR sensors.

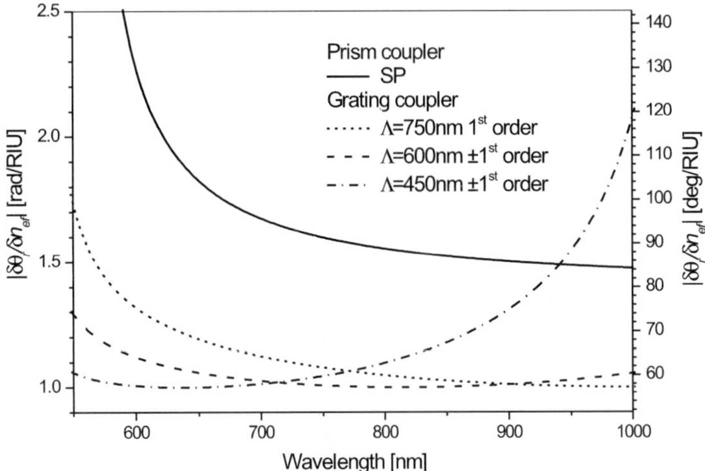

Fig. 5 Instrumental contribution to sensitivity $\delta\theta_r/\delta n_{ef}$ as a function of wavelength for SPR sensors with angular modulation and prism coupler or grating coupler and three different grating periods. Prism-based sensor configuration: BK7 glass prism, gold film, and a non-dispersive dielectric (refractive index 1.32). Grating-based sensor configuration: a non-dispersive dielectric (refractive index 1.32) and gold grating

As follows from Fig. 5, the sensitivity $(\delta\theta_r/\delta n_{ef})_{prism}$ increases with a decreasing wavelength following the wavelength dependence of the effective index of the surface plasmon and the coupling angle (see Chap. 1 of this volume [1]). The rapid increase of $(\delta\theta_r/\delta n_{ef})_{prism}$ at short wavelengths is associated with the effective index of a surface plasmon n_{ef} approaching the refractive index of the prism n_p (and angle of incidence approaching 90 deg). The instrumental contribution to sensitivity $(\delta\theta_r/\delta n_{ef})_{grating}$ exhibits a minimum that corresponds to the normal incidence ($\lambda = \Lambda n_{ef}$) and increases both towards long and short wavelengths.

For SPR sensors with angular modulation using coupled surface plasmons on a thin metal film (instead of conventional surface plasmons on a metal–dielectric interface), the instrumental contribution to sensitivity $\delta\theta_r/\delta n_{ef}$ can also be calculated from Eqs. 12 and 13.

As follows from Eq. 12, the instrumental contribution $(\delta\theta_r/\delta n_{ef})_{prism}$ is determined by the difference between the refractive index of the prism and effective index of the surface plasmon. Therefore, the highest sensitivity can be obtained using an antisymmetric surface plasmon (which, however, exists only within a range of wavelengths at which its effective index is smaller than the refractive index of prism) (Fig. 6). SPR sensors employing a symmetric surface plasmon are less sensitive than those using conventional plasmon at a metal–dielectric interface (Fig. 6). The sensitivity

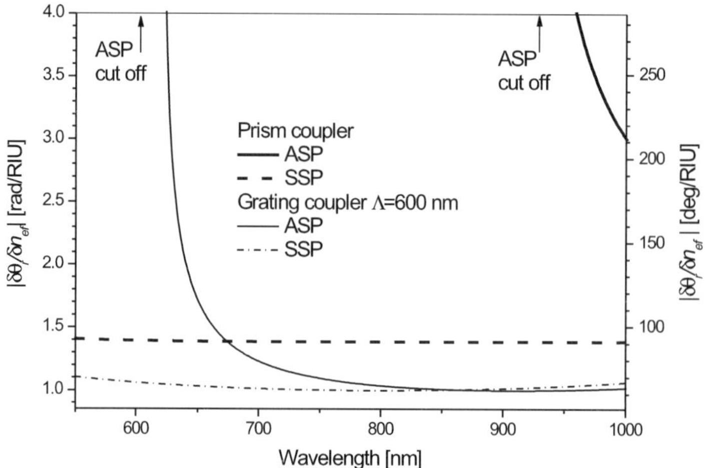

Fig. 6 Instrumental contribution to sensitivity $\delta\theta_r/\delta n_{ef}$ as a function of wavelength for an SPR sensor with angular modulation which employs symmetric (SSP) and antisymmetric (ASP) surface plasmons excited on a thin gold film using prism or grating coupler. Prism-based sensor configuration: BK7 glass prism, buffer layer (refractive index 1.32), gold film (thickness 20 nm), and a non-dispersive dielectric (refractive index 1.32).Grating-based sensor configuration: a non-dispersive dielectric (refractive index 1.32) and grating (grating period 600 nm) supporting a gold film (thickness 20 nm) and a buffer layer (refractive index 1.32)

$\left(\delta\theta_r/\delta n_{ef}\right)_{grating}$ of grating-based SPR sensors using symmetric and antisymmetric surface plasmons follows basically the same trend, however, the cut-off for the antisymmetric mode $\left(n_{ef} = (\lambda m/\Lambda + 1)\right)$ is shifted to shorter wavelengths.

Once the instrumental contribution to sensitivity $\delta\theta_r/\delta n_{ef}$ of an SPR sensor has been determined, the sensitivity to refractive index can be calculated as follows:

$$S_{RI} = \left|\frac{\delta\theta_r}{\delta n_{ef}}\right| \frac{\delta n_{ef}}{\delta n}, \tag{14}$$

where the term $\delta n_{ef}/\delta n$ describes the sensitivity of the effective index of a surface plasmon to refractive index and depends on the details of the distribution of the refractive index change. In the following section we calculate the sensor sensitivity to bulk refractive index change to illustrate this process and to provide a practical means for evaluation of the sensitivity of various configurations of SPR sensors.

The sensitivity of angular modulation-based SPR sensors to bulk refractive index (herein denoted as $\left(S_\theta\right)_{prism}$ and $\left(S_\theta\right)_{grating}$ for SPR sensors using prism and grating couplers, respectively) can be derived from Eqs. 12 and 13, and the equation for $\delta n_{ef}/\delta n$ obtained using the perturbation theory (Eq. 58

in Chap. 1 of this volume [1]) [13]:

$$\left(S_\theta\right)_{\text{prism}} = \frac{\varepsilon'_m \sqrt{-\varepsilon'_m}}{\left(\varepsilon'_m + n^2\right)\sqrt{\varepsilon'_m\left(n^2 - n_{\text{p}}^2\right) - n^2 n_{\text{p}}^2}}, \tag{15}$$

$$\left(S_\theta\right)_{\text{grating}} = \frac{\varepsilon'_m \sqrt{-\varepsilon'_m}}{\left(\varepsilon'_m + n^2\right)\sqrt{\left(1 - \frac{m^2\lambda^2}{\Lambda^2}\right)\left(n^2 + \varepsilon'_m\right) - n^2\varepsilon'_m + 2n\frac{|m|\lambda}{\Lambda}\sqrt{\varepsilon'_m\left(n^2 + \varepsilon'_m\right)}}}, \tag{16}$$

where ε'_m is the real part of the permittivity of metal. Bulk refractive index sensitivities of model SPR sensors with angular modulation and prism and grating coupler were calculated using (Eq. 15) and (Eq. 16) and are shown in Fig. 7. As the sensitivity of the effective index of the surface plasmon to bulk refractive index decreases slowly with an increasing wavelength, the behavior of the bulk refractive index sensitivity basically follows the instrumental contribution $\delta\theta_{\text{r}}/\delta n_{\text{ef}}$. The bulk refractive index sensitivity of SPR sensors using symmetric or antisymmetric surface plasmons is also dominated by the instrumental contribution $\delta\theta_{\text{r}}/\delta n_{\text{ef}}$ (Fig. 8).

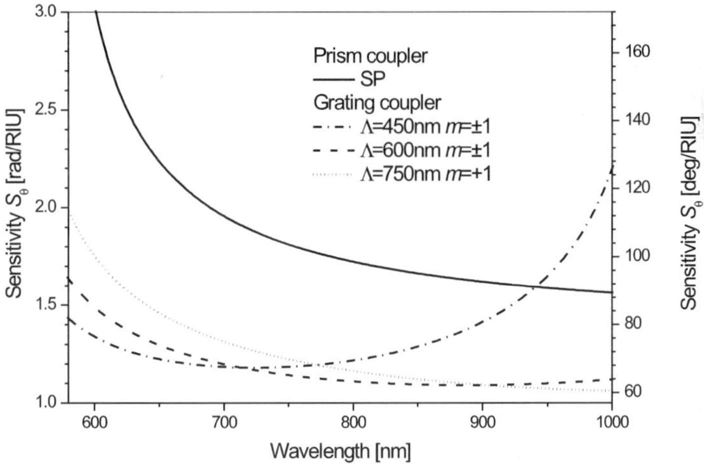

Fig. 7 Bulk refractive index sensitivity as a function of wavelength for SPR sensors with angular modulation and prism coupler or grating coupler and three different grating periods. Prism-based sensor configuration: BK7 glass prism, gold film, and a non-dispersive dielectric (refractive index1.32). Grating-based sensor configuration: a non-dispersive dielectric (refractive index 1.32) and gold grating

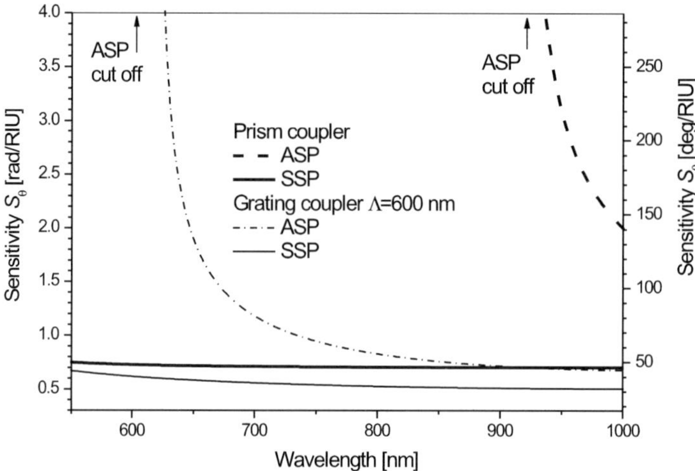

Fig. 8 Bulk refractive index sensitivity as a function of wavelength for an SPR sensor with angular modulation which employs symmetric (SSP) and antisymmetric (ASP) surface plasmons excited on a thin gold film using prism or grating coupler. Prism-based sensor configuration: BK7 glass prism, buffer layer (refractive index 1.32), gold film (thickness 20 nm), and a non-dispersive dielectric (refractive index 1.32). Grating-based sensor configuration: a non-dispersive dielectric (refractive index 1.32) and grating (grating period 600 nm) supporting a gold film (thickness 20 nm) and a buffer layer (refractive index 1.32)

4.1.2 Sensitivity of SPR Sensors with Wavelength Modulation

In SPR sensors with angular modulation, the sensor output is the coupling wavelength λ_r and therefore the instrumental contribution to sensor sensitivity S_{RI1} is equal to $\delta\lambda_r/\delta n_{ef}$. By differentiating the coupling conditions for the prism, grating, and waveguide coupler (see Chap. 1 of this volume [1]) in n_{ef} and λ, we obtain:

$$\left(\frac{\delta\lambda_r}{\delta n_{ef}}\right)_{\text{prism}} = \frac{1}{\frac{dn_p}{d\lambda}\frac{n_{ef}}{n_p} - \frac{dn_{ef}}{d\lambda}}, \tag{17}$$

$$\left(\frac{\delta\lambda_r}{\delta n_{ef}}\right)_{\text{grating}} = \frac{1}{\frac{|m|}{\Lambda} - \frac{dn_{ef}}{d\lambda}}, \tag{18}$$

$$\left(\frac{\delta\lambda_r}{\delta n_{ef}}\right)_{\text{waveguide}} = \frac{1}{\frac{dn_{wg}}{d\lambda} - \frac{dn_{ef}}{d\lambda}}, \tag{19}$$

where the derivatives $dn_{ef}/d\lambda$, $dn_p/d\lambda$ and $dn_{wg}/d\lambda$ describe the dispersion of the effective index of a surface plasmon, dispersion of the coupling prism, and chromatic dispersion of the waveguide, respectively. The dispersion of glasses constituting prism couplers is usually much smaller than the dispersion of surface plasmons on a metal–dielectric interface. For the wavelengths and typical materials used in waveguide-based SPR sensors, the chromatic dispersion $\mathrm{d}n_{\mathrm{wg}}/\mathrm{d}\lambda$ is to a large extent determined by the material dispersion of the waveguide. The wavelength dependence of the instrumental contribution to sensitivity $\delta\lambda_{\mathrm{r}}/\delta n_{\mathrm{ef}}$ for model prism and grating-based SPR sensors, calculated using Eqs. 17 and 18, is shown in Fig. 9.

The instrumental contribution to sensitivity of SPR sensors using prism couplers $\left(\delta\lambda_{\mathrm{r}}/\delta n_{\mathrm{ef}}\right)_{\mathrm{prism}}$ is primarily determined by the second term in the denominator of Eq. 17 $\mathrm{d}n_{\mathrm{ef}}/\mathrm{d}\lambda$ and therefore $\left(\delta\lambda_{\mathrm{r}}/\delta n_{\mathrm{ef}}\right)_{\mathrm{prism}}$ increases with increasing wavelength. The instrumental contribution to sensitivity $\left(\delta\lambda_{\mathrm{r}}/\delta n_{\mathrm{ef}}\right)_{\mathrm{grating}}$ increases with an increasing wavelength and levels off as the term m/Λ dominates over $\mathrm{d}n_{\mathrm{ef}}/\mathrm{d}\lambda$. The instrumental contribution to sensitivity of SPR sensors with prism coupler $\left(\delta\lambda_{\mathrm{r}}/\delta n_{\mathrm{ef}}\right)_{\mathrm{prism}}$ is typically larger by an order of magnitude than that for a grating coupler $\left(\delta\lambda_{\mathrm{r}}/\delta n_{\mathrm{ef}}\right)_{\mathrm{grating}}$.

For SPR sensors with wavelength modulation using coupled surface plasmons on a thin metal film (instead of conventional surface plasmons on a metal–dielectric interface), the instrumental contribution to sensitivity $\delta\lambda_{\mathrm{r}}/\delta n_{\mathrm{ef}}$ can be also calculated from Eqs. 17 and 18 (Fig. 10). The wavelength dependence of the instrumental contribution to sensitivity $\delta\lambda_{\mathrm{r}}/\delta n_{\mathrm{ef}}$ for model prism and grating-based SPR sensors employing symmetric and antisymmet-

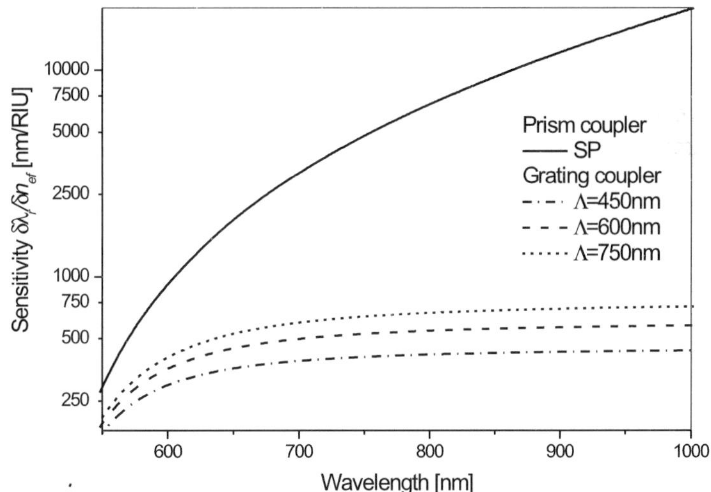

Fig. 9 Instrumental contribution to sensitivity $\delta\lambda_{\mathrm{r}}/\delta n_{\mathrm{ef}}$ as a function of wavelength for SPR sensors with wavelength modulation and prism coupler or grating coupler and three different grating periods. Prism-based sensor configuration: BK7 glass prism, gold film, and a non-dispersive dielectric (refractive index 1.32). Grating-based sensor configuration: a non-dispersive dielectric (refractive index 1.32) and gold grating

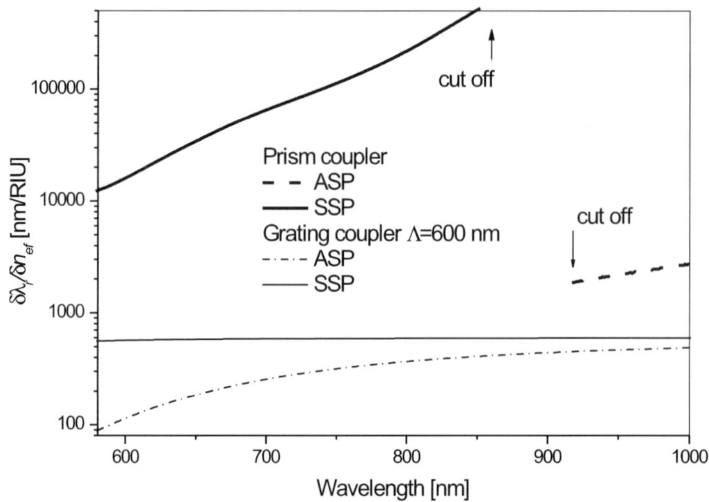

Fig. 10 Instrumental contribution to sensitivity $\delta\lambda_r/\delta n_{ef}$ as a function of wavelength for an SPR sensor with wavelength modulation which employs symmetric (SSP) and antisymmetric (ASP) surface plasmons excited on a thin gold film using prism or grating coupler. Prism-based sensor configuration: BK7 glass prism, buffer layer (refractive index 1.32), gold film (thickness 20 nm), and a non-dispersive dielectric (refractive index 1.32). Grating-based sensor configuration: a non-dispersive dielectric (refractive index 1.32) and grating (grating period 600 nm) supporting a gold film (thickness 20 nm) and a buffer layer (refractive index 1.32)

ric surface plasmons, calculated using Eqs. 17 and 18, is shown in Fig. 10. The instrumental contribution to sensitivity $(\delta\lambda_r/\delta n_{ef})_{prism}$ for the symmetric surface plasmon increases until the denominator in Eq. 17 reaches zero [14] and is larger by an order of magnitude than that for the conventional surface plasmon on a metal–dielectric interface, or for the antisymmetric surface plasmon. The instrumental contribution to sensitivity of SPR sensors using grating couplers $(\delta\lambda_r/\delta n_{ef})_{grating}$ and symmetric or antisymmetric surface plasmon increases with increasing wavelength and is significantly smaller than $(\delta\lambda_r/\delta n_{ef})_{prism}$ and comparable with $(\delta\lambda_r/\delta n_{ef})_{grating}$ values obtained using conventional surface plasmons on gold–dielectric interface.

Once the instrumental contribution to sensitivity $\delta\lambda_r/\delta n_{ef}$ of an SPR sensor has been determined, the sensitivity to refractive index can be calculated as follows:

$$S_{RI} = \frac{\delta\lambda_r}{\delta n_{ef}} \frac{\delta n_{ef}}{\delta n} , \tag{20}$$

where the term $\delta n_{ef}/\delta n$ describes the sensitivity of the effective index of a surface plasmon to refractive index and depends on the details of the distribution of the refractive index change. The spectral sensitivity of SPR sensors based on prism and grating couplers and wavelength modulation for conven-

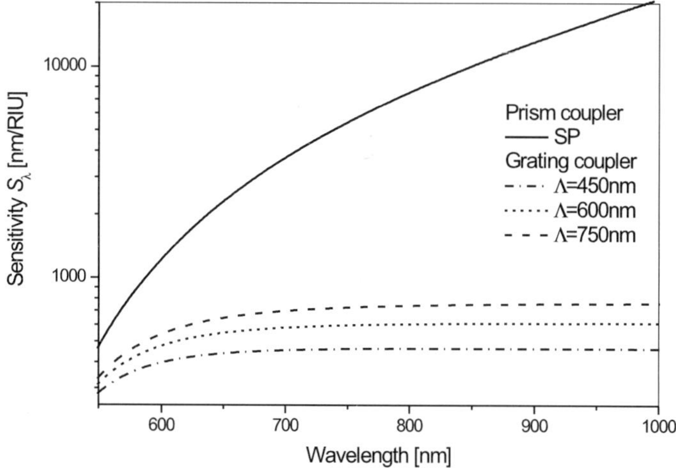

Fig. 11 Bulk refractive index sensitivity as a function of wavelength for SPR sensors with wavelength modulation and prism coupler or grating coupler and three different grating periods. Prism-based sensor configuration: BK7 glass prism, gold film, and a non-dispersive dielectric (refractive index 1.32). Grating-based sensor configuration: a non-dispersive dielectric (refractive index 1.32) and gold grating

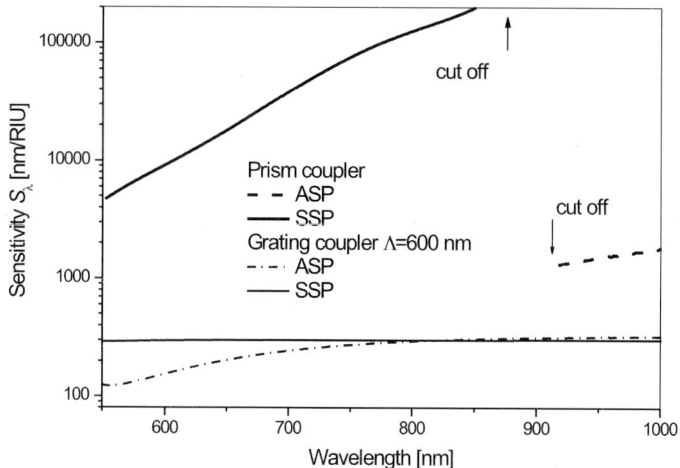

Fig. 12 Bulk refractive index sensitivity as a function of wavelength for SPR sensors with wavelength modulation employing symmetric (SSP) and antisymmetric (ASP) surface plasmons excited on a thin gold film using prism or grating coupler. Prism-based sensor configuration: BK7 glass prism, buffer layer (refractive index 1.32), gold film (thickness 20 nm), and a non-dispersive dielectric (refractive index 1.32). Grating-based sensor configuration: a non-dispersive dielectric (refractive index 1.32) and grating (grating period 600 nm) supporting a gold film (thickness 20 nm) and a buffer layer (refractive index 1.32)

tional surface plasmons on a gold–dielectric interface and coupled surface plasmons on a thin gold film is shown in Figs. 11 and 12.

Analytical expressions for sensitivity of SPR sensors with wavelength modulation can be derived from Eq. 20 using Eqs. 17 and 18 ($\delta n_{\mathrm{ef}}/\delta n$ is taken from Eq. 58 in Chap. 1 of this volume [1]) or directly from the coupling conditions [13]:

$$\left(S_\lambda\right)_{\text{prism}} = \frac{\varepsilon_m'^2}{-\frac{n^3}{2}\frac{\mathrm{d}\varepsilon_m'}{\mathrm{d}\lambda} + \varepsilon_m'\left(n^2+\varepsilon_m'\right)\frac{n}{n_\mathrm{p}}\frac{\mathrm{d}n_\mathrm{p}}{\mathrm{d}\lambda}}, \tag{21}$$

$$\left(S_\lambda\right)_{\text{grating}} = \frac{\varepsilon_m'^2}{-\frac{n^3}{2}\frac{\mathrm{d}\varepsilon_m'}{\mathrm{d}\lambda} + \frac{|m|}{\Lambda}\left(n^2+\varepsilon_m'\right)\sqrt{\varepsilon_m'\left(n^2+\varepsilon_m'\right)}}, \tag{22}$$

where $\mathrm{d}\varepsilon_m'/\mathrm{d}\lambda$ and $\mathrm{d}n_\mathrm{p}/\mathrm{d}\lambda$ represent the material dispersion of the metal and prism respectively.

4.1.3 Sensitivity of SPR Sensors with Intensity Modulation

In SPR sensors with intensity modulation, the sensor output is the intensity (which is proportional to reflectivity in prism or grating-based SPR sensor) and therefore the instrumental contribution to sensor sensitivity S_{RI1} can be written as $\delta R/\delta n_{\mathrm{ef}}$, where R denotes reflectivity. In order to derive analytical expressions for the instrumental contributions to sensor sensitivity, the reflectivity is assumes to follow the Lorentzian shape (Eq. 78 in Chap. 1 of this volume [1]):

$$R \doteq 1 - \frac{4\gamma_i\gamma_{\mathrm{rad}}}{\Delta_{\mathrm{ef}}^2 + (\gamma_i+\gamma_{\mathrm{rad}})^2}, \tag{23}$$

where γ_i and γ_{rad} denote the attenuation coefficients of surface plasmons due to the absorption and radiation, respectively [15] and:

$$\Delta_{\mathrm{ef}} = n_\mathrm{p}\sin\theta - n_{\mathrm{ef}}, \tag{24}$$

for prism coupler. Using Eq. 23, the sensitivity of reflectivity to effective refractive index can be expressed as follows:

$$\frac{\partial R}{\partial n_{\mathrm{ef}}} = \frac{-8\Delta_{\mathrm{ef}}\gamma_i\gamma_{\mathrm{rad}}}{\left[\Delta_{\mathrm{ef}}^2 + (\gamma_i+\gamma_{\mathrm{rad}})^2\right]^2}. \tag{25}$$

Yeatman et al. showed that the maximum slope of the reflectivity occurs when:

$$\Delta_{\mathrm{ef}} = \pm\frac{(\gamma_i+\gamma_{\mathrm{rad}})}{\sqrt{3}} \tag{26}$$

and that the maximum slope can be obtained when $\gamma_{\mathrm{rad}} = \gamma_i/2$ [16]. This condition reduces the depth of the resonant dip, but the decrease of its width results in the increase of the reflectivity dependence slope. Under these conditions, the maximum instrumental contribution to sensitivity can be expressed as:

$$\frac{\delta R}{\delta n_{\mathrm{ef}}} = \left(\frac{\partial R}{\partial n_{\mathrm{ef}}}\right)_{\max} = \frac{2\sqrt{3}}{9\gamma_i} . \tag{27}$$

The wavelength dependence of the instrumental contribution to sensitivity $\delta R/\delta n_{\mathrm{ef}}$ for model prism-based SPR sensors calculated using Eq. 27 is shown in Fig. 13.

The bulk refractive index sensitivity can be calculated from the instrumental sensitivity in a similar fashion as in previous sections for the angular and wavelength modulation. The sensitivity to bulk refractive index can be expressed as:

$$S_{\mathrm{RI}} = \frac{\delta R}{\delta n_{\mathrm{ef}}} \frac{\delta n_{\mathrm{ef}}}{\delta n} . \tag{28}$$

The maximum sensitivity of intensity modulation-based SPR sensors to bulk refractive index $\left(S_{\mathrm{I}}\right)_{\max}$ can be derived from Eq. 28 (where $\delta n_{\mathrm{ef}}/\delta n$ is from

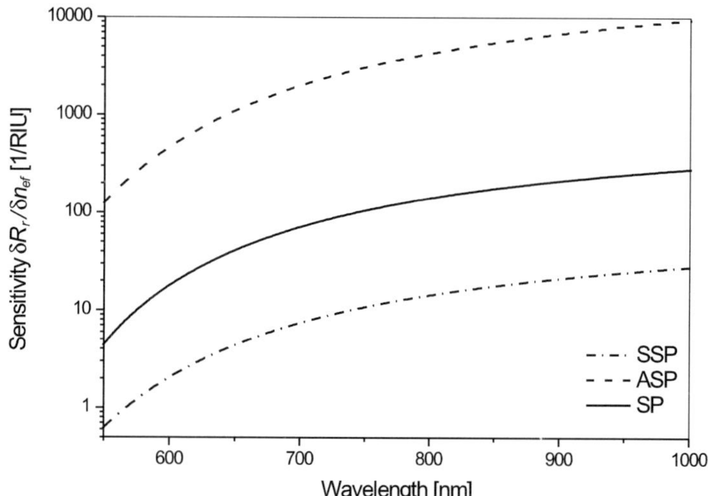

Fig. 13 Instrumental contribution to sensitivity $\delta R/\delta n_{\mathrm{ef}}$ as a function of wavelength for SPR sensors with intensity modulation and a prism coupler exciting a conventional surface plasmon (SP) at the interface of gold and a non-dispersive dielectric (refractive index 1.32) and coupled symmetric (SSP) and antisymmetric (ASP) surface plasmons on a thin gold film (thickness 20 nm) surrounded by two identical non-dispersive dielectrics (refractive index 1.32)

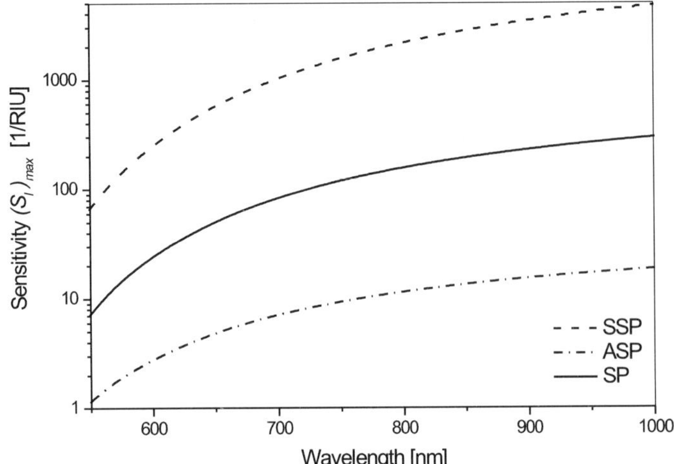

Fig. 14 The maximum bulk refractive index sensitivity as a function of wavelength for SPR sensors with intensity modulation and a prism coupler exciting a conventional surface plasmon (SP) at the interface of gold and a non-dispersive dielectric (refractive index 1.32) and coupled symmetric (SSP) and antisymmetric (ASP) surface plasmons on a thin gold film (thickness 20 nm) surrounded by two identical non-dispersive dielectrics (refractive index 1.32)

Eq. 58 in Chap. 1 of this volume [1]) as follows:

$$\left(S_{\mathrm{I}}\right)_{\max} = \frac{4\sqrt{3}}{9} \frac{\varepsilon_m'^2}{\varepsilon_m'' n^3} , \tag{29}$$

where ε_m' and ε_m'' are the real and imaginary part of the metal permittivity, respectively. Figure 14 shows the maximum sensitivity calculated using Eq. 28 for selected model SPR sensor configurations. As follows from Fig. 14, the bulk refractive index sensitivity of intensity-based sensors increases with increasing wavelength. This behavior is caused by the attenuation coefficient of surface plasmons, which decreases with increasing wavelength.

4.2 Resolution

Resolution of SPR sensors defines the smallest change in the bulk refractive index that produces a detectable change in the sensor output. The magnitude of sensor output change that can be detected depends on the level of uncertainty of the output, the output noise.

The noise of sensor output originates in the optical system and readout electronics of an SPR sensor instrument. Dominant sources of noise are fluctuations in the light intensity emitted by the light source, statistical properties of light (shot noise), and noise in conversion of light intensity into photoelectrons by the detector and supporting circuitry.

The noise in the intensity of light emitted by the light source is proportional to the intensity and therefore its standard deviation σ can be given as $\sigma_L = \sigma_L^{rel} I$, where σ_L^{rel} is relative (intensity-independent) standard deviation and I denotes the measured light intensity.

The shot noise is associated with random arrival of photons on a detector and corresponding random production of photoelectrons. Conventional light sources produce photon flux that obeys Poisson statistics, which produces a shot noise σ_S directly proportional to the square root of the detected light intensity: $\sigma_S = \sigma_S^{rel}\sqrt{I}$, where σ_S^{rel} is relative (intensity-independent) standard deviation [17]. The detector noise consists of several contributions that originate mostly in temperature generated photoelectrons and the detector electronic circuitry, and its standard deviation σ_D is independent of the detected light intensity. The resulting noise of the measured light intensity σ_I is a statistical superimposition of all the noise components and can therefore be expressed as:

$$\sigma_I(I) = \sqrt{I^2\left(\sigma_L^{rel}\right)^2 + I\left(\sigma_S^{rel}\right)^2 + \sigma_D^2}\,. \tag{30}$$

In SPR sensors based on wavelength or angular modulations, multiple intensities corresponding to different wavelengths or angles of incidence are acquired. This results in series of wavelength or angular spectra in time ($I_1, I_2, I_3, \ldots, I_N$). These spectra are mathematically processed to generate the sensor output. In the first phase of data processing, the spectra are usually averaged (to reduce noise) and normalized (to eliminate effects of unequal angular or wavelength light distribution). The averaging either involves averaging of time series of intensity from the same detector (time averaging) or averaging of intensities from multiple detectors (e.g., of a 2D array) measured at a single time (spatial averaging). As in the time domain, all the noise contributions behave independently, the time averaging of N spectra reduces the noise of each intensity in the spectrum as follows:

$$\sigma_I^{tN} = \sqrt{I^2\frac{\left(\sigma_L^{rel}\right)^2}{N} + \frac{I\left(\sigma_S^{rel}\right)^2}{N} + \frac{\sigma_D^2}{N}} = \frac{\sigma_I}{\sqrt{N}}\,. \tag{31}$$

Spatial domain averaging can be applied when the sensor signal from N different detectors is averaged. In wavelength or angular modulation it is usually applied to spectra that are measured in several rows of a 2D detector [18, 19]. In sensors based on intensity modulation such as SPR imaging, the averaged area corresponds to one measuring channel [20–22]. In spatial averaging, the light fluctuations affect all the intensities measured by different detectors in the same way and therefore the light source noise is not reduced by the spatial averaging:

$$\sigma_{\mathrm{I}}^{\mathrm{sN}} = \sqrt{I^2 \left(\sigma_{\mathrm{L}}^{\mathrm{rel}}\right)^2 + \frac{I\left(\sigma_{\mathrm{S}}^{\mathrm{rel}}\right)^2}{N} + \frac{\sigma_{\mathrm{D}}^2}{N}} > \frac{\sigma_{\mathrm{I}}}{\sqrt{N}} \,. \tag{32}$$

Therefore, especially in SPR sensors with intensity modulation, the light source noise can dominate over the shot noise and detector noise and needs to be reduced by other means.

The averaged and normalized spectra are translated into the sensor output by an appropriate data processing algorithm. Numerous methods for calculation of sensor output have been used in SPR sensors such as the centroid method [23, 24], polynomial fitting followed with the analytical calculation of the polynomial minimum [25, 26], and optimal linear data analysis [27]. As can be shown by a more general analysis, the noise in angular or wavelength spectra transforms to the noise in the sensor output in a similar fashion for the most common algorithms [28]. Therefore, we shall use the centroid method as a model data processing algorithm to illustrate the propagation of noise into the sensor output.

The centroid method uses a simple algorithm which finds the geometric center of the portion of the SPR dip under a certain threshold. Although the geometric center does not necessarily coincide with the minimum of the spectrum, as SPR sensing usually relies on relative measurements, the offset of the geometric center does not affect the final measurements. The centroid is calculated as follows:

$$Y_{\mathrm{C}} = \frac{\sum\limits_j x_j \left(I_{\mathrm{thresh}} - I_j\right)}{\sum\limits_j \left(I_{\mathrm{thresh}} - I_j\right)} \,, \tag{33}$$

where x_j represent the spectral positions of the contributing intensities I_j and I_{thresh} denotes the threshold value. If the noise of intensities detected by individual detectors can be treated as independent, the resulting standard deviation of calculated dip position (sensors output noise) σ_{so} can be derived from the noise of individual intensities $\sigma(I_j)$ as:

$$\sigma_{\mathrm{so}}^2 = \sum_{j=1}^{N} \left(\frac{\partial Y_{\mathrm{C}}}{\partial I_j}\right)^2 \sigma^2\left(I_j\right) , \tag{34}$$

where N is the number of involved intensities, and $\partial Y_{\mathrm{C}}/\partial I_j$ denotes the incremental contribution factors to the noise of dip position Y_{C} (sensor output) from each detector [29]. If we assume that the portion of the SPR dip used by the centroid algorithm follows the Lorentzian profile (see Chap. 1 in this volume [1]), that the optimum threshold level is at the half of the SPR dip depth [29], and substitute Y_C from Eq. 33 in Eq. 34, we obtain:

$$\sigma_{SO} \doteq \frac{1.75}{Nd}\sqrt{\sum_{j=1}^{N}(Y_C - x_j)^2\sigma^2(I_j)}\,, \tag{35}$$

where Y_C is the centroid position, N is the number of intensities below the threshold, and d is the depth of the dip (difference of intensities between the dip minimum and threshold). For the three types of noise discussed above: (a) the additive noise independent of the intensity, (b) the noise proportional to the square root of intensity (shot noise) and (c) the noise proportional to the intensity, Eq. 35 yields:

$$\sigma_{SO} = K\frac{\sigma_{th}}{d}\cdot\frac{w}{\sqrt{N}}\,, \tag{36}$$

where σ_{th} is the intensity noise at threshold, d is the depth of the dip (difference of intensities between the dip minimum and threshold), w is the width of the dip (at the threshold) and $K = K_1 = 0.50$ for additive noise, $K = K_2 = 0.43$ for the shot noise, and $K = K_3 = 0.38$ for the noise proportional to the intensity. If the intensity noise is superimposition of the three types of noise with weights g_1, g_2, and g_3 ($g_1 + g_2 + g_3 = 1$), the coefficient K can be calculated as $K = \sqrt{g_1K_1^2 + g_2K_2^2 + g_3K_3^2}$. This yields a refractive index uncertainty σ_{RI}:

$$\sigma_{RI} = \frac{\sigma_{SO}}{S_{RI}} = K\frac{\sigma_{th}}{d\sqrt{N}}\cdot\frac{w}{S_{RI}}\,, \tag{37}$$

where S_{RI} denotes the bulk refractive index sensitivity of the sensor. The width of the SPR dip is directly correlated with its sensitivity and it can be shown that the ratio w/S_{RI} depends only weakly on the choice of coupler and modulation.

Equation 36 indicates that the sensor output noise is linearly dependent on the noise of measured signal and inverse linearly dependent on the depth of the SPR dip. If the number of measured intensities N is proportional to the investigated spectrum width, the sensor output noise depends on the square root of the width.

Using Eq. 37, the ultimate resolution of an SPR sensor can be predicted. Figure 15 shows theoretical resolutions calculated using Eq. 37 for SPR sensors employing three different types of detectors (a linear CCD detector, 2D CCD, and PDA detector). The following parameters were used for the simulations of the SPR sensor using a linear CCD detector: shot noise of 0.6% at the saturation level of the detector [29], and time averaging over 500 spectra (limited by typical readout and exposure times), which yields $(\sigma_{th}/d) = 4 \times 10^{-4}$ (threshold is set to half of the detector dynamical range). The following parameters were used for the SPR sensor using a 2D detec-

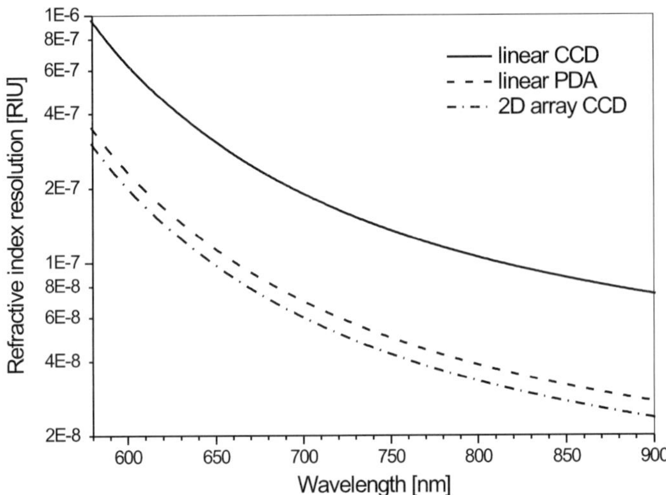

Fig. 15 Modeled resolution of SPR configuration with linear CCD, 2D CCD, and large area photodiode array (PDA) as a function of wavelength

tor: shot noise of 0.6%, spatial averaging over 100 detector lines [30], and time averaging over 50 spectra, which yields $(\sigma_{\text{th}}/\text{d}) = 1 \times 10^{-4}$ (threshold at half of the detector dynamical range). The following parameters were used for the simulations of the SPR sensor using a linear photodiode array (PDA) detector: the noise is dominantly additive ($g_1 = 0.8$, $g_2 = 0.2$, $K = 0.49$) and its SD relative to the threshold set to half of the detector dynamical range is 0.1% [29]. The time averaging over 50 spectra is supposed, yielding $(\sigma_{\text{th}}/\text{d}) = 1 \times 10^{-4}$. Furthermore, it is assumed that the width of the SPR dip covers 300 pixels ($N = 300$), which is close to the configuration analyzed in [29–31].

Figure 15 suggests that SPR sensors with large area detectors such as PDA or 2D array CCD can potentially achieve a better resolution compared to systems using linear CCD detectors. This comparison is related to the fact that more light is measured with large area detectors in the same period of time, which results in the reduction of the shot noise.

5 Summary

Surface plasmon resonance (SPR) sensors are optical sensing devices that take advantage of the sensitivity of a special type of electromagnetic field, a surface plasmon, to changes in refractive index. SPR sensors can be classified based on the method for optical excitation of surface plasmons and the measured characteristic of the light wave interacting with the surface plasmon.

SPR sensors directly measure refractive index. In conjunction with appropriate biorecognition elements, they can be used as affinity biosensors allowing detection of the capture of analyte molecules by biorecognition elements immobilized on the sensor surface. The ability of SPR sensors to perform measurements is described by the performance characteristics, of which the most important are the sensitivity, resolution, accuracy, reproducibility, and limit of detection. The sensitivity and resolution are primarily determined by the properties of the optical system of the SPR sensor and can be linked to specific design parameters.

References

1. Homola J (2006) Electromagnetic Theory of Surface Plasmons. In: Homola J (ed) Surface Plasmon Resonance Based Sensors. Springer Ser Chem Sens Biosens, vol 4. Springer, Berlin Heidelberg New York (in this volume)
2. Homola J (2003) Anal Bioanal Chem 377:528
3. Matsubara K, Kawata S, Minami S (1988) Appl Spect 42:1375
4. Zhang LM, Uttamchandan D (1988) Electron Lett 24:1469
5. Nylander C, Liedberg B, Lind T (1982) Sensor Actuator 3:79
6. Brockman JM, Nelson BP, Corn RM (2000) Ann Rev Phys Chem 51:41
7. Kruchinin AA, Vlasov YG (1996) Sensor Actuator B Chem 30:77
8. Ho C, Robinson A, Miller D, Davis M (2005) Sensors 5:4
9. Tumolo T, Angnes L, Baptista MS (2004) Anal Biochem 333:273
10. de Feijter JA, Benjamins J, Veer FA (1978) Biopolymers 17:1759
11. Thomsen V, Schatzlein D, Mercuro D (2003) Spectroscopy 18:112
12. Currie LA (1997) Chemomet Intel Lab Sys 37:151
13. Homola J, Koudela I, Yee SS (1999) Sensor Actuator B Chem 54:16
14. Nenninger GG, Tobiška P, Homola J, Yee SS (2001) Sensor Actuator B Chem 74:145
15. Johansen K, Arwin H, Lundstrom I, Liedberg B (2000) Rev Sci Inst 71:3530
16. Yeatman EM (1996) Biosens Bioelectron 11:635
17. Drake AW (1967) Fundamentals of applied probability theory. McGraw-Hill, New York
18. O'Brien MJ, Perez-Luna VH, Brueck SRJ, Lopez GP (2001) Biosens Bioelectron 16:97
19. Kawazumi H, Gobi KV, Ogino K, Maeda H, Miura N (2005) Sensor Actuator B Chem 108:791
20. Piliarik M, Vaisocherová H, Homola J (2005) Biosens Bioelectron 20:2104
21. Berger CEH, Beumer TAM, Kooyman RPH, Greve J (1998) Anal Chem 70:703
22. Fu E, Chinowsky T, Foley J, Weinstein J, Yager P (2004) Rev Sci Inst 75:2300
23. Kukanskis K, Elkind J, Melendez J, Murphy T, Miller G, Garner H (1999) Anal Biochem 274:7
24. Goddard NJ, Pollardknight D, Maule CH (1994) Analyst 119:583
25. Sjölander S, Urbanitzky C (1991) Anal Chem 63:2338
26. Stenberg E, Persson B, Roos H, Urbaniczky C (1991) J Colloid Interface Sci 143:513
27. Chinowsky TM, Jung LS, Yee SS (1999) Sensor Actuator B Chem 54:89
28. Tobiška P, Homola J (2005) Sensor Actuator B Chem 107:162
29. Nenninger GG, Piliarik M, Homola J (2002) Meas Sci Technol 13:2038
30. Thirstrup C, Zong W, Borre M, Neff H, Pedersen HC, Holzhueter G (2004) Sensor Actuator B Chem 100:298
31. Dostálek J, Homola J, Miler M (2005) Sensor Actuator B Chem 107:154

Springer Ser Chem Sens Biosens (2006) 4: 69–91
DOI 10.1007/5346_015

Published online: 25 May 2006

Molecular Interactions in SPR Sensors

Josef Štěpánek[1] (✉) · Hana Vaisocherová[2] · Marek Piliarik[2]

[1]Faculty of Mathematics and Physics, Charles University, Prague, Czech Republic
stepjos@karlov.mff.cuni.cz

[2]Institute of Radio Engineering and Electronics, Prague, Czech Republic

Keywords Association · Diffusion · Dissociation · Equilibrium constant · Flow cell · Mass transport · Rate constants · Reaction kinetics · Sensor · Surface plasmon resonance

Abbreviations

A	Analyte (free reagent in solution)
B	Complex of analyte and receptor
D	Diffusion coefficient (coefficient of translational diffusion)
Da	Damköhler number
h	Flow cell height
k_a	Association rate constant
k_d	Dissociation rate constant
K	Equilibrium association constant (binding affinity)
l	Flow cell length
N_A	Avogadro's number
PDE	Fundamental equation of analyte transport inside flow cell (partial differential equation)
Pe	Peclét number
R	Receptor (binding target for the analyte immobilized at the sensor surface)
RU	Units of the SPR sensor response (resonance units)
Re	Reynolds number
w	Flow cell width
[X]	Molar concentration of X
x	Space coordinate in direction of the analyte flow

y Space coordinate in direction perpendicular to the sensor surface
α Free analyte concentration ($\equiv [A]$)
α_0 Injected analyte concentration
β Surface concentration of receptor (moles per square area)
γ Surface concentration of complex (indexed for various types of complexes when necessary)
ξ Sensor response (in RU)
ξ_S Standard sensor response (in RU) corresponding to all receptor sites bound to analyte in 1 : 1 ratio

1 Introduction

Binding and/or unbinding of biomolecules at the active surface of an SPR biosensor is controlled by various mechanisms that result in variety of temporal profiles of the SPR biosensor response and in dependence on microenvironmental conditions. The determination of binding kinetics provides important new information about interacting molecules. This is commonly considered one of the greatest advantages of the SPR biosensor technique. Although in ideal cases an appropriate kinetic model of molecular interaction is able to completely describe the SPR biosensor response, in reality the influence of hydrodynamic conditions often has to be taken into account [1]. This chapter is devoted to molecular interaction models that correspond to the processes most frequently encountered at SPR biosensor surfaces. It also deals with hydrodynamic effects and their exact or approximate mathematical description.

2 Interaction Models

To quantitatively analyze the sensor response to interactions between the studied biomolecule (analyte) and the surface bound receptors, it is necessary to employ a relevant mathematical model. The core part of the model is a kinetic equation that describes how the temporal amounts of formed/dissociated complexes depend on the momentary local concentrations of the free analyte and the free binding sites of the receptors.

SPR biosensor experiments measure only relative changes in the molecular mass attached to the sensor surface from the beginning of the interaction being studied. The response ξ is then directly proportional to the concentration of the bound analyte (conditions that guarantee a linear sensor response are assumed throughout the chapter). In the case of a single type of analyte binding to the receptors in a 1 : 1 stoichiometric ratio, the response is proportional to the concentration of the formed complexes:

$$\xi = \text{const}\, M_\mathrm{A} \gamma\,, \tag{1}$$

where M_A is the mass of the analyte molecule and γ is the surface concentration of the formed complexes. It can be shown that for sufficiently high analyte concentrations, the sensor response will eventually reach its maximum value, which corresponds to all of the receptors being occupied. This response does not change measurably with further increases in the analyte concentration. Considering that the maximum possible response for the 1 : 1 stoichiometry is given by:

$$\xi_\mathrm{S} = \text{const}\, M_\mathrm{A} \beta\,, \tag{2}$$

where β is the surface concentration of receptors, it is widely useful to characterize the sensor response by its normalized value:

$$\xi/\xi_\mathrm{S} = \gamma/\beta\,. \tag{3}$$

2.1 Pseudo First-Order Kinetics

Whenever we deal with analyte binding to receptors fixed at a sensor surface, the second order reaction model represents the basis of its description. This model concerns the situation when two partners, A and R, form a single complex AR. This can be, for instance, binding of an antigen to an antibody, docking of a substrate to an enzyme with a single binding pocket, or duplex formation by two complementary chains of nucleic acid. In the case of interactions at the sensor surface, we have to distinguish between the immobilized receptor R and the analyte A present in the solution. Two processes are considered by the model: (1) the association process whereby A and R bind to each other and create the immobilized complex AR and (2) the dissociation process whereby the complex AR dissociate into two parts, A and R. These processes are symbolized by:

$$\mathrm{A} + \mathrm{R} \rightarrow \mathrm{AR} \quad \text{and} \quad \mathrm{AR} \rightarrow \mathrm{A} + \mathrm{R}\,. \tag{4}$$

For the association, it is essential that A and R are in close proximity, i.e., their distance must be shorter than a critical radius. If this condition is satisfied, there is a certain probability that within a unit time interval A and R will form a complex. For a set of given environmental conditions (temperature, pressure, solvent properties) this probability is the same for all neighboring pairs of A and R, provided we do not consider microscopic conditions such as their mutual orientation or their instantaneous speeds of translation and rotation. For a given receptor the probability that any molecule of analyte appears within the critical distance is proportional to the concentration of A. The total number of associations per time interval in a particular region is proportional to the total number of receptors involved, because they all can create a complex with the same probability. As a result, we obtain a relationship between the amount of the complexes γ formed per unit time, the instantaneous concentration of the free analyte $[A] = \alpha$, and the concentration of free receptors $\beta - \gamma$:

$$\frac{\mathrm{d}\gamma_\mathrm{a}}{\mathrm{d}t} = k_\mathrm{a}\alpha\left(\beta - \gamma\right) , \tag{5}$$

where k_a is a constant that characterizes the chemical reaction in the sense that it is independent of time and of the reactants concentrations. It is called the association or forward rate constant.

On the other hand, for each complex there is certain probability that within a unit time interval it will dissociate into A and R separated by a distance larger than the critical radius. This probability is the same for all complexes at the given conditions. The dissociation leads to a decrease of the complex concentration proportional to its instantaneous value:

$$\frac{\mathrm{d}\gamma_\mathrm{d}}{\mathrm{d}t} = - k_\mathrm{d}\gamma , \tag{6}$$

where k_d is called the dissociation or reverse rate constant. In a real system, both the association and dissociation processes occur simultaneously. It can be symbolically expressed as:

$$\mathrm{A} + \mathrm{R} \underset{k_\mathrm{d}}{\overset{k_\mathrm{a}}{\rightleftharpoons}} \mathrm{AR} . \tag{7}$$

The time dependence of the total complex concentration is then described by the summed effects of both processes:

$$\frac{\mathrm{d}\gamma}{\mathrm{d}t} = \frac{\mathrm{d}\gamma_\mathrm{a}}{\mathrm{d}t} + \frac{\mathrm{d}\gamma_\mathrm{d}}{\mathrm{d}t} = k_\mathrm{a}\alpha\left(\beta - \gamma\right) - k_\mathrm{d}\gamma . \tag{8}$$

Both quantities β and γ must be expressed in the same kind of local density. In the case of a solution phase reaction, we would understand them as molar concentrations, i.e., number of moles per unit volume. For receptors fixed on the sensor surface it is more straightforward to define them as surface concentrations, i.e., number of moles per unit area.

The solution of Eq. 8 depends strongly on how the concentration of the free analyte α is controlled. In the case of an active sensor surface surrounded by a solvent occupying certain closed volume V, the analyte can be injected as a highly concentrated solution [1, 2]. In the ideal case of a perfectly mixed solution, the effect of the injection can be described as an immediate jump in the analyte concentration from zero to a certain starting value α_0. During the consequent process the free analyte will be consumed by association with the receptor, while the sum of the free and bound analyte will be kept constant:

$$\alpha V + \gamma S = \alpha_0 V = \text{const} , \tag{9}$$

where S is the sensor area. The temporary change in the complex concentration is then proportional to a quadratic polynomial of its instantaneous value (Fig. 1):

$$\frac{d\gamma}{dt} = k_a \left(\alpha_0 - \frac{S}{V}\gamma \right) (\beta - \gamma) - k_d \gamma \,. \tag{10}$$

At longer times, the solution of Eq. 10 converges to an equilibrium state $\left(\frac{d\gamma}{dt} = 0 \right)$, which is characterized by the well-known equation:

$$K = \frac{k_a}{k_d} = \frac{\gamma_{eq}}{\left(\alpha_0 - \gamma_{eq} S/V \right) \left(\beta - \gamma_{eq} \right)} \,, \tag{11}$$

where K is the equilibrium (association) constant. Sometimes it is also referred to as the binding affinity. A sense of Eq. 11 is demonstrated in Fig. 1: changes in the association rate influence how fast both the concentration of the complexes and that of the free analyte come to equilibrium. Note that their equilibrium values are not changed, because the equilibrium constant is fixed.

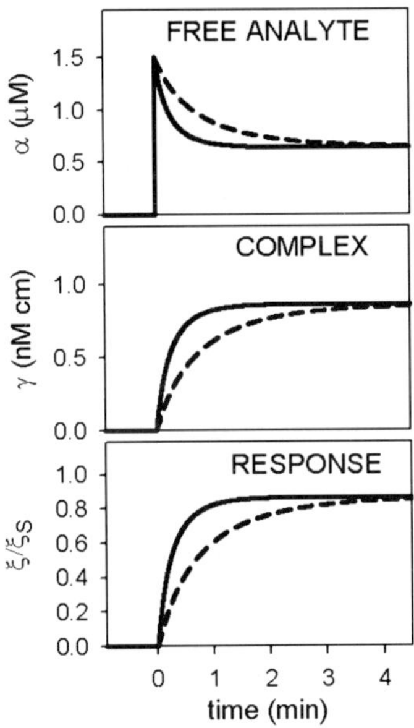

Fig. 1 Analyte-to-receptor binding after the analyte injection, according to the model of the second order reaction in closed volume (Eq. 10). Parameters: $\alpha_0 = 1.5\ \mu$M, $\beta = 10^{-9}$ M cm, $S = 1$ cm^2, $V = 1\ \mu$L, $K = 10^7$ M^{-1}. *Solid line* $k_a = 4.5 \times 10^4$ M^{-1} s^{-1}, *dashed line* $k_a = 1.5 \times 10^4$ M^{-1} s^{-1}

Very often the number of molecules of analyte in volume V is much higher than the amount of the receptors at the surface S. In this case, the term $\gamma S/V$ in Eq. 10 can be neglected and we obtain:

$$\frac{\mathrm{d}\gamma}{\mathrm{d}t} = k_\mathrm{a}\alpha_0 \left(\beta - \gamma\right) - k_\mathrm{d}\gamma \,. \tag{12}$$

Equation 12, originally derived by Langmuir for interactions at a surface in contact with reactants in solution, is formally identical with the equation describing a first-order reaction in solution. It is therefore usually referred to as pseudo first-order kinetics. Its solution is a single exponential function with an asymptote corresponding to the equilibrium fulfilling equation:

$$K = \frac{k_\mathrm{a}}{k_\mathrm{d}} = \frac{\gamma_\mathrm{eq}}{\alpha_0 \left(\beta - \gamma_\mathrm{eq}\right)} \,. \tag{13}$$

Pseudo first-order kinetics is also typical for sensors that employ flow cells, where the free analyte concentration is primarily controlled by flowing a solution through the cell. In this case, the free analyte concentration can be either increased stepwise or decreased stepwise. As a result, an SPR sensorgram usually consists of two stages: an association stage that begins with the stepwise increase of the free analyte concentration to a constant value α_0, followed by a dissociation stage where the free analyte concentration is stepped down to zero. An ideal SPR response corresponding to this experiment (pseudo first-order kinetics) is shown in Fig. 2.

After a sufficiently long time, the association and the dissociation rates become practically equal and a dynamic equilibrium state is achieved. The

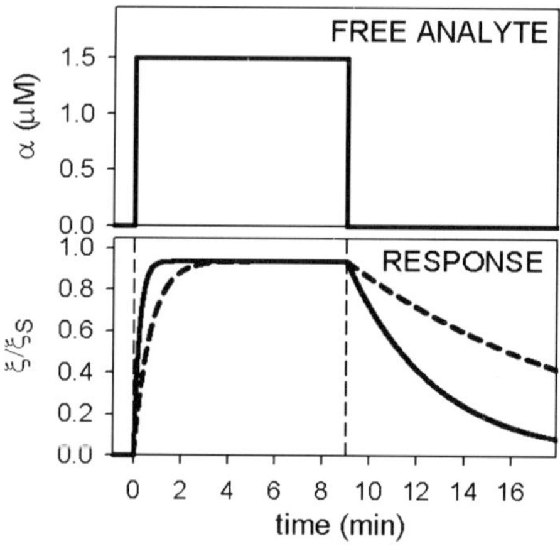

Fig. 2 Ideal flow-cell sensorgram according to the model of pseudo first-order reaction (Eq. 12). Parameters: $\alpha_0 = 1.5\,\mu\mathrm{M}$, $\beta = 1\,\mathrm{nM\,cm}$, $K = 10^7\,\mathrm{M}^{-1}$. *Solid line* $k_\mathrm{a} = 4.5 \times 10^4\,\mathrm{M}^{-1}\,\mathrm{s}^{-1}$, *dashed line* $k_\mathrm{a} = 1.5 \times 10^4\,\mathrm{M}^{-1}\,\mathrm{s}^{-1}$. *Vertical dashed lines* indicate beginning of the association and the dissociation stage

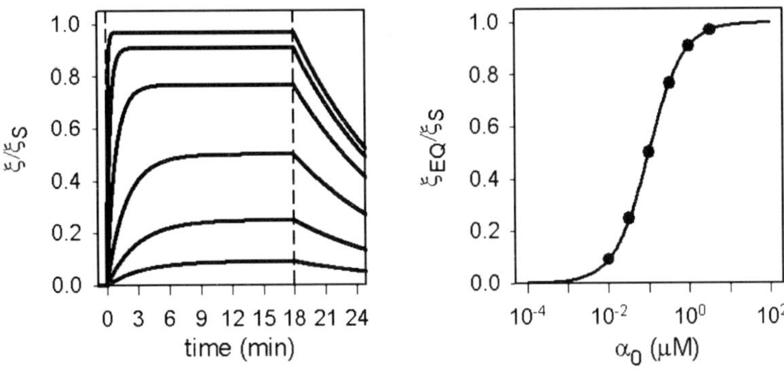

Fig. 3 Set of sensorgrams (model) suitable for equilibrium analysis (*left*) and binding isotherm with indicated equilibrium sensorgram results (*right*). Model of pseudo first-order reaction. Parameters: $\beta = 1$ nM cm, $K = 10^7\ \mathrm{M}^{-1}$, $k_a = 4.5 \times 10^4\ \mathrm{M}^{-1}\ \mathrm{s}^{-1}$

equilibrium association constant K [3–7] can be determined by measuring the dependence of the sensor's equilibrium response on the injected analyte concentration (binding isotherm). For pseudo first-order kinetics the binding isotherm (Fig. 3) is given by:

$$\frac{\xi_{EQ}}{\xi_S} = \frac{K\alpha_0}{(1 + K\alpha_0)} \, . \tag{14}$$

An advantage of equilibrium analysis is that, in contrast to the other parts of the sensorgram, the equilibrium phase of the association curve is not affected by mass transport (see below).

2.2 Other Kinetic Models

In reality, the processes in the active sensor layer may be more complicated and the sensor response will be a superposition of several parallel or consecutive reactions. We will present some kinetic models that correspond to more complex molecular interactions at the sensor surface.

Zero order reactions following the initial binding are usually interpreted as conformational changes of the AR complex. Once the conformation is changed, the complex cannot dissociate unless it transforms back into its original state. This additional reaction can slow down the kinetics. The model, first presented in [8], has been applied in a few studies of complex biomolecular systems where the analyte binding may substantially change the physicochemical properties of the receptor, such as the interaction of angiotensin II with a receptor at a lipid membrane [9] or the interactions of sulfated polysaccharides with immobilized enzyme targets [10]. The reaction scheme of this two-state model is:

$$A + R \underset{k_{d1}}{\overset{k_{a1}}{\rightleftharpoons}} AR \underset{k_{d2}}{\overset{k_{a2}}{\rightleftharpoons}} AR^* . \tag{15}$$

Assigning γ_1 and γ_2 to the concentrations of the complex in particular states:

$$\gamma_1 = [AR] , \quad \gamma_2 = \left[AR^*\right] , \tag{16}$$

the corresponding kinetic equations that account for the relationships between both types of complexes and the free receptor sites can be written as:

$$\begin{aligned} \frac{\partial \gamma_2}{\partial t} &= k_{a2}\gamma_1 - k_{d2}\gamma_2 \\ \frac{\partial \gamma_1}{\partial t} &= k_{a1}\alpha_0\left(\beta - \gamma_1 - \gamma_2\right) - \frac{\partial \gamma_2}{\partial t} \\ &= k_{a1}\alpha_0\left(\beta - \gamma_1 - \gamma_2\right) - k_{d1}\gamma_1 - k_{a2}\gamma_1 + k_{d2}\gamma_2 . \end{aligned} \tag{17}$$

As the conformational change does not influence the mass of the complex, the sensor response will be:

$$\xi/\xi_S = \left(\gamma_1 + \gamma_2\right)/\beta . \tag{18}$$

Models of parallel pseudo first-order reactions consider the case when two interactions with different rate constants proceed simultaneously. Such situations can be attributed to different kinds of receptor sites or to different states of the analyte [8, 11]. In the first case the model can describe heterogeneity of the sensor surface; the second may concern a macromolecular analyte that can be present in various conformations, protonation states, etc. Besides two sets of rate constants, the models also require specification of proportion p between the two fractions of the receptor or analyte. For the model considering two kinds of receptors, the following equations are obtained:

$$A + R_1 \underset{k_{d1}}{\overset{k_{a1}}{\rightleftharpoons}} AR_1 \quad A + R_2 \underset{k_{d2}}{\overset{k_{a2}}{\rightleftharpoons}} AR_2 \tag{19}$$

$$\begin{aligned} &\beta_1 = [R_1] = p\beta \quad \beta_2 = [R_2] = \left(1 - p\right)\beta \quad \gamma_1 = [AR_1] \quad \gamma_2 = [AR_2] \\ &\frac{d\gamma_1}{dt} = k_{a1}\alpha_0\left(\beta_1 - \gamma_1\right) - k_{d1}\gamma_1 \quad \frac{d\gamma_2}{dt} = k_{a2}\alpha_0\left(\beta_2 - \gamma_2\right) - k_{d2}\gamma_2 \\ &\xi/\xi_S = \left(\gamma_1 + \gamma_2\right)/\beta . \end{aligned} \tag{20}$$

Results of this model are illustrated in Fig. 4. For the case with two states of the analyte, the equations are analogous to the previous ones except that the

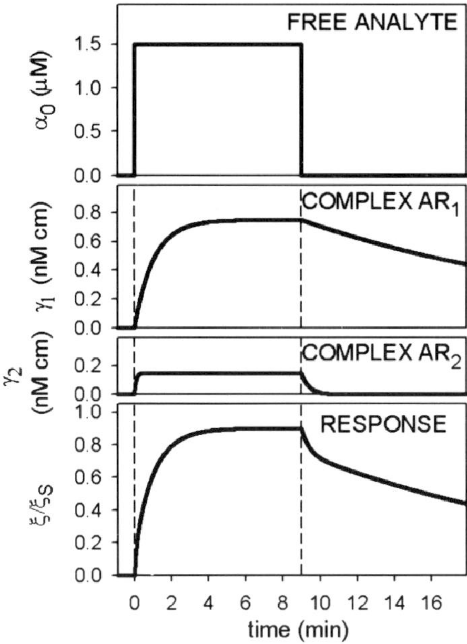

Fig. 4 Kinetics and sensorgram according to the model of two parallel pseudo first-order reactions attributed to two kinds of receptors (Eq. 19). Parameters: $\alpha_0 = 1.5\,\mu\text{M}$, $\beta = 1$ nM cm, $p = 0.8$, $k_{a1} = 10^4\,\text{M}^{-1}\,\text{s}^{-1}$, $k_{d1} = 0.001\,\text{s}^{-1}$, $k_{a2} = 8 \times 10^4\,\text{M}^{-1}\,\text{s}^{-1}$, $k_{d2} = 0.04\,\text{s}^{-1}$

effect of competition for the receptor sites must be included:

$$\text{A}_1 + \text{R} \underset{k_{d1}}{\overset{k_{a1}}{\rightleftharpoons}} \text{A}_1\text{R} \quad \text{A}_2 + \text{R} \underset{k_{d2}}{\overset{k_{a2}}{\rightleftharpoons}} \text{A}_2\text{R} \tag{21}$$

$$\alpha_1 = [\text{A}_1] = p\alpha_0 \quad \alpha_2 = [\text{A}_2] = (1-p)\,\alpha_0 \quad \gamma_1 = [\text{A}_1\text{R}] \quad \gamma_2 = [\text{A}_2\text{R}]$$

$$\frac{\text{d}\gamma_1}{\text{d}t} = k_{a1}\alpha_1(\beta - \gamma_1 - \gamma_2) - k_{d1}\gamma_1 \quad \frac{\text{d}\gamma_2}{\text{d}t} = k_{a2}\alpha_2(\beta - \gamma_1 - \gamma_2) - k_{d2}\gamma_2 \tag{22}$$

$$\xi/\xi_S = (\gamma_1 + \gamma_2)/\beta\,.$$

Equations 22 can also be employed in the case of two different analytes, although the last relationship for calculating the sensor response must be modified to account for the different masses of the analytes.

Multivalent receptor binding is a case when a single receptor molecule can bind more than one molecule of analyte. Multivalent binding capacity is a frequent feature of many biomolecular systems, for instance antibodies. Another example is the formation of triplexes by oligonucleotides. If a purine oligonucleotide is fixed at the sensor surface as a receptor, a complementary oligonucleotide can bind to it and to create a duplex. In special cases, another oligonucleotide molecule may bind to the duplex and form a triplex.

This situation involves two successive reactions, each occurring at a unique binding site. The corresponding kinetic equations are:

$$\mathrm{A} + \mathrm{R} \underset{k_{\mathrm{d1}}}{\overset{k_{\mathrm{a1}}}{\rightleftharpoons}} \mathrm{AR} \quad \mathrm{AR} + \mathrm{A} \underset{k_{\mathrm{d2}}}{\overset{k_{\mathrm{a2}}}{\rightleftharpoons}} \mathrm{ARA} \tag{23}$$

$$\gamma_1 = [\mathrm{AR}] \quad \gamma_2 = [\mathrm{ARA}]$$

$$\frac{\partial \gamma_2}{\partial t} = k_{\mathrm{a2}}\alpha_0\gamma_1 - k_{\mathrm{d2}}\gamma_2 \tag{24}$$

$$\begin{aligned}\frac{\partial \gamma_1}{\partial t} &= k_{\mathrm{a1}}\alpha_0\left(\beta - \gamma_1 - \gamma_2\right) - k_{\mathrm{d1}}\gamma_1 - \frac{\partial \gamma_2}{\partial t} \\ &= k_{\mathrm{a1}}\alpha_0\left(\beta - \gamma_1 - \gamma_2\right) - k_{\mathrm{d1}}\gamma_1 - k_{\mathrm{a2}}\alpha_0\gamma_1 + k_{\mathrm{d2}}\gamma_2\end{aligned}$$

$$\xi/\xi_{\mathrm{S}} = \left(\gamma_1 + 2\gamma_2\right)/\beta\,.$$

Here the standard sensor response is assumed to be the case when all receptors are bound in 1 : 1 complexes (duplexes). That is why the relative response can exceed 1, as is seen in Fig. 5.

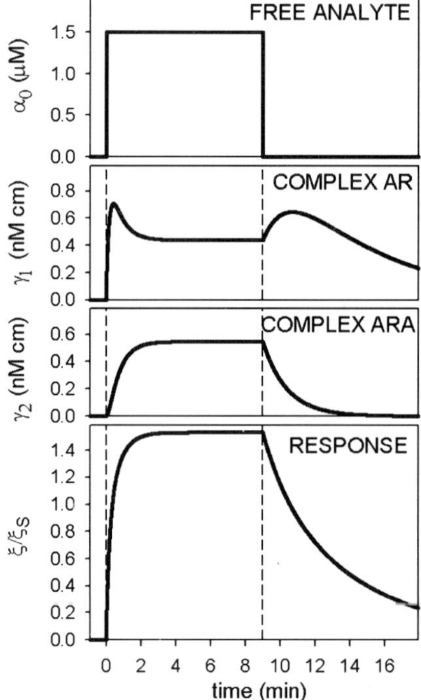

Fig. 5 Kinetics and sensorgram according to the model of consecutive two binding reactions in case of bivalent receptor (Eq. 23). Parameters: $\alpha_0 = 1.5\,\mu\mathrm{M}$, $\beta = 1\,\mathrm{nM\,cm}$, $k_{\mathrm{a1}} = 6 \times 10^4\,\mathrm{M}^{-1}\,\mathrm{s}^{-1}$, $k_{\mathrm{d1}} = 0.003\,\mathrm{s}^{-1}$, $k_{\mathrm{a2}} = 10^4\,\mathrm{M}^{-1}\,\mathrm{s}^{-1}$, $k_{\mathrm{d2}} = 0.012\,\mathrm{s}^{-1}$

Binding of a multivalent analyte occurs when a single analyte molecule can simultaneously occupy more than one receptor molecule. This case does not mirror the previous one, because the resulting analyte/receptor interaction strongly depends on the receptor distribution on the sensor surface. For sufficiently sparse receptor spacing, only a single binding mode is available despite the number of analyte binding sites – once the analyte is caught by a receptor, it is isolated from other distant receptors. Increasing the receptor density increases the probability of forming a receptor pattern that allows an analyte to bind multiple receptors. In [12] the authors introduced the concept of dividing the active sensor layer into spheres with a radius equal to the functional distance between the two binding sites of the analyte. Using Poisson statistics they estimated the portion p of those receptors (R_C) that were at least by two inside one sphere. Remaining receptors (R_S) were expected to be single within a sphere. The kinetic model considered simple 1 : 1 binding on R_S receptors and consecutive binding on R_C receptors:

$$A + R_S \underset{k_d}{\overset{k_a}{\rightleftharpoons}} AR_S \quad A + R_C \underset{k_d}{\overset{k_a}{\rightleftharpoons}} AR_C \quad AR_C + R_C \underset{k_d}{\overset{k_a}{\rightleftharpoons}} AR_CR_C \,. \tag{25}$$

Assuming the same rate constants for both binding sites on the analyte, the following set of equations is obtained:

$$\beta_1 = [R_S] = (1-p)\,\beta \quad \beta_2 = [R_C] = p\beta$$
$$\gamma_1 = [AR_S] \quad \gamma_2 = [AR_C] \quad \gamma_3 = [AR_CR_C]$$
$$\frac{\partial \gamma_1}{\partial t} = 2k_a\alpha_0\left(\beta_1 - \gamma_1\right) - k_d\gamma_1 \quad \frac{\partial \gamma_2}{\partial t} = 2k_a\alpha_0\left(\beta_2 - \gamma_2 - 2\gamma_3\right) - k_d\gamma_2 - \frac{\partial \gamma_3}{\partial t} \tag{26}$$
$$\frac{\partial \gamma_3}{\partial t} = k_a\gamma_2\frac{\beta_2 - \gamma_2 - 2\gamma_3}{\beta_2}\frac{1}{V_{sp}N_A} - 2k_d\gamma_3$$
$$\xi/\xi_S = \left(\gamma_1 + \gamma_2 + 2\gamma_3\right)/\beta \,.$$

Note that a factor of 2 appears in the kinetic equations to account for the doubled probability because of two binding sites on the analyte. In the equation for $\frac{\partial \gamma_3}{\partial t}$, the fraction $\frac{\beta_2 - \gamma_2 - 2\gamma_3}{\beta_2}$ is the probability that there is a free receptor inside the sphere where the AR_C complex occurs. The second fraction, $\frac{1}{V_{sp}N_A}$, where V_{sp} is the sphere volume and N_A is Avogadro's number, represents the concentration of the available analyte – one molecule in the V_{sp} sphere. It has been demonstrated in [12] that this model fits experimental data substantially better than a solvent kinetic model of multiple binding, which does not respect the fixed positions of the receptors.

At the end of this section it is worth mentioning that besides SPR studies where the analyte binding to the receptor is the only running interaction, competitive SPR biosensor experiments with two concurrent interactions, i.e., analyte-immobilized receptor and analyte-another ligand in solution, can also be performed. Proper kinetic models for competitive SPR studies should be developed based on the appropriate kinetic equations for the particular interactions, using an approach analogous to the aforementioned cases.

2.3 Thermodynamic Context of Equilibrium and Kinetic Constants

The equilibrium association constant K is directly related to the change of the molar Gibbs energy attributed to complex formation ΔG^0:

$$\Delta G^0 = -RT \ln \left(K_\mathrm{a} C^0 \right) , \tag{27}$$

where R is the universal gas constant, T is the absolute temperature, and C^0 is a standard concentration – as a rule its value is taken as 1 M. The basic temperature dependence of ΔG is given by the van't Hoff equation:

$$\Delta G = \Delta H - T\Delta S , \tag{28}$$

where ΔH and ΔS are the changes of enthalpy and of entropy. If they are both temperature independent, a plot of $\ln \left(K_\mathrm{a} C^0 \right)$ versus $1/T$ (van't Hoff plot) should be linear. The ΔH and ΔS values can be determined directly from the graph; more precise is a least square fit of Eqs. 27 and 28.

The simple van't Hoff equation (Eq. 28) is not completely correct if the complex formation results in a change of the specific heat capacity ΔC_p, in which case neither ΔH nor ΔS are exactly independent of temperature. A more precise form of the van't Hoff equation is [13]:

$$\Delta G \left(T \right) = \Delta H_\mathrm{To} - T\Delta S_\mathrm{To} + \Delta C_\mathrm{p} \left(T - T_0 \right) + \Delta C_\mathrm{p} T \ln \left(\frac{T}{T_0} \right) , \tag{29}$$

where T_0 is a reference temperature. To obtain reliable values of ΔH_To, ΔS_To, and ΔC_p (T_0 is defined, usually $T_0 = 298.15\,\mathrm{K}$, i.e., 25 °C), precise data over a wider range of temperatures are necessary for the fit. Estimation of any of the thermodynamic parameters from another experiment is very helpful.

The temperature dependence of k_a and k_d is usually characterized by means of activation energy ($E_\mathrm{a}^\mathrm{act}$ and $E_\mathrm{d}^\mathrm{act}$) according to the Arrhenius equation:

$$\ln k = \ln P - \frac{E^\mathrm{act}}{RT} , \tag{30}$$

where P is a constant known as the pre-exponential factor. The activation energy is assumed to be a measure of the amount of thermal energy required for binding or dissociation. Because $E_\mathrm{a}^\mathrm{act}$ and $E_\mathrm{d}^\mathrm{act}$ can be considered as activation enthalpies, the reaction enthalpy can be calculated from the relationship:

$$\Delta H = E_\mathrm{a}^\mathrm{act} - E_\mathrm{d}^\mathrm{act} . \tag{31}$$

An unusually high E^{act} value indicates that binding and/or dissociation requires the surmounting of high potential energy barriers, suggesting that conformational rearrangements are required.

When possible, the kinetic rate constants determined using SPR sensors have been compared to those obtained in bulk solution using other methods. Good agreement was obtained only in some cases. For instance, it has been reported [14] that when a study of the interactions between small inhibitor molecules and immobilized proteins was carefully designed, performed, and analyzed, very good agreement with the bulk data was achieved.

The basic formula for the association rate constant is given by Debye–Smoluchowski theory:

$$k_a = 4\pi\varphi\varepsilon r\left(D_A + D_R\right)N_A/1000\,, \tag{32}$$

where φ is a steric interaction factor, ε is an electrostatic interaction factor, r is an interaction radius. D_A and D_R are translation diffusion coefficients of the analyte and the receptor molecule.

Let us consider the term of the translation diffusion. The diffusion coefficient D expresses the ability of a molecule to change its position in solution due to chaotic translation motion. Basic evaluation of the diffusion coefficient can be obtained from the Stokes formula for a sphere in a fluid:

$$D = \frac{k_B T}{6\pi a\eta}\,, \tag{33}$$

where k_B is the Boltzmann constant, T is the absolute temperature, η is the viscosity of the fluid and a is the radius of the sphere approximating the molecule size. Therefore, the diffusion coefficient decreases strongly with increasing size of the molecule. In contrast to the case of both interaction partners in solution, the translation diffusion of the receptor is limited when it is immobilized at the sensor surface. The value inside the parentheses in Eq. 32 may then be reduced and approximated as close to the D_A term alone. The final effect of immobilization on the translation diffusion term would depend on the ratio between D_A and D_R. If the receptor is a large molecule like protein and the analyte is a small molecule like the inhibitors used in the experiments reported in [14], then $D_A \approx (D_A + D_R)$ and the association and dissociation rate constants may be very close for both the SPR biosensor and for reactions in the bulk. On the other hand, a small receptor interacting with a large analyte may be characterized by rate constants significantly different from those measured in the bulk.

To conclude this section we summarize that, in general, the kinetic rate constants obtained from SPR sensors may not agree with those obtained in solution. The SPR technique seems to be better suited to performing comparative studies of molecules according their affinity and other interaction characteristics. However, improvements in the precision of SPR measurements and of their theoretical description may soon lead to new approaches for extracting fundamental information about biomolecular interactions using SPR sensors with receptors with varying degrees of restricted mobility.

3
Mass Transport Effects

In SPR biosensors, the rate of biomolecular interactions at the surface depends on the free analyte transport toward (association stage) or away from (dissociation stage) the active zone. The stepwise free analyte concentration changes discussed above are only an idealization, because the free analyte transport is always limited. The influence of analyte transport on the reactions at the sensor surface is given by comparing the transport throughput to the kinetic rates. Slow analyte transport causes a decrease in its concentration when it is consumed during the association phase and an increase when it is produced during the dissociation phase. As a result, both reactions are slowed down.

This effect is illustrated in Fig. 6 where the kinetics of a simple pseudo first-order reaction are calculated assuming that analyte transport is propor-

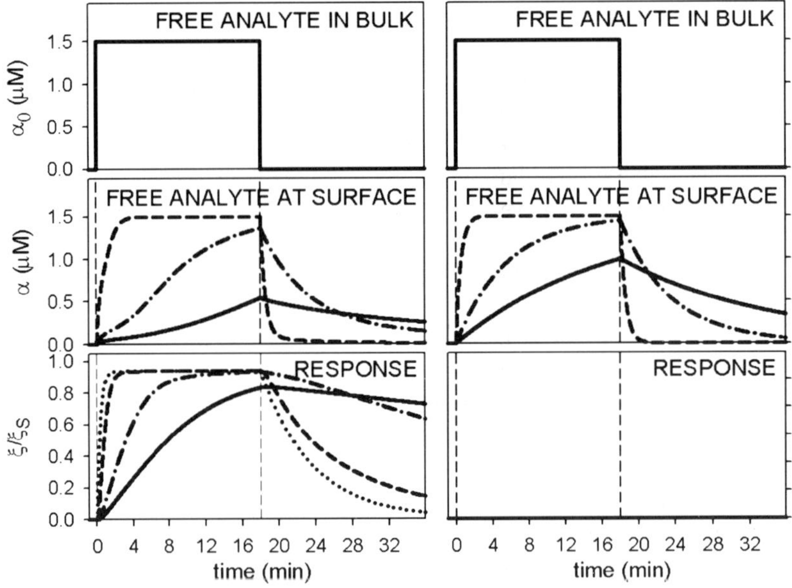

Fig. 6 Free analyte concentrations in the active layer and sensorgrams for the pseudo first-order reaction and two-compartment model of the analyte transport. Parameters: α_0 = 1.5 μM, $\beta = 1$ nM cm. *Left* $k_a = 0.03\ \mathrm{M}^{-1}\ \mathrm{s}^{-1}$, $k_d = 0.003\ \mathrm{s}^{-1}$; *right* no binding of analyte to receptor. Rate constant of the analyte diffusion flux (Eq. 32) $k_M/h_{layer} = 3 \times 10^{-5}\ \mathrm{s}^{-1}\ \mathrm{cm}^{-1}$ (*dashed line*), $3 \times 10^{-6}\ \mathrm{s}^{-1}\ \mathrm{cm}^{-1}$ (*dash-and-dot*), and $10^{-6}\ \mathrm{s}^{-1}\ \mathrm{cm}^{-1}$ (*solid*). *Dotted line* no limitations of the analyte transport

tional to the concentration difference between the bulk analyte solution and the active sensor layer (two-compartment model, discussed in greater detail later). In addition to the effect on analyte binding kinetics, the figure also clearly illustrates that the reaction between the receptor and analyte influences, i.e., significantly reduces the free analyte concentration. If the reaction does not occur (right-hand figures), the free analyte concentration reaches its equilibrium value more rapidly.

3.1 Analyte Transport in a Flow Cell

The flow cell shape is typically rectangular with its length l (dimension along the flow) and width w (dimension perpendicular to the flow and parallel to the sensor surface) in the range 10^{-0}–10^{-2} cm, and a substantially lower height h (dimension perpendicular to the sensor surface) measuring 10^{-2}–10^{-3} cm. Flow characteristics can be described by the Reynolds number:

$$\mathrm{Re} = \frac{\rho \Phi}{\eta h}, \tag{34}$$

where ρ and η are the density and viscosity of the fluid, and Φ is the flow rate (volume of fluid passing through the cell per unit time interval). The flow is expected to be laminar (without turbulence) if $\mathrm{Re} < 2100$ [15]. For water at 20 °C, $\mathrm{Re} = (\Phi/h) \cdot 0.998\ \mathrm{mm^2\,s^{-1}}$. Considering typical flow cell dimensions and flow rates, the Reynolds number does not exceed several hundreds. The distance between the active sensor surface and both the inlet and outlet is as a rule far enough that the laminar flow profile is fully developed in the active sensor region [16]. The velocity profile is therefore considered as constant over the sensor active zone.

Let us introduce spatial coordinates in the flow-cell interior: x in direction of the length, y in direction of the height, and z in direction of the width. The magnitude of the velocity (its direction is uniformly parallel to the x-axis) depends mainly on y. The velocity profile is parabolic, with the maximum velocity $v_{\max}$ at the mid-point of the cell height and zero velocity at the cell walls. In contrast, the velocity dependence on z is negligible (except for the regions very close to the cell walls, which are sufficiently far from the active region) [17]. The total fluid flux through the flow cell can thus be obtained by integrating over the y coordinate from zero to h. This provides a relation between $v_{\max}$, the cell dimensions, and Φ:

$$v_{\max} = \frac{3}{2}\frac{\Phi}{hw}. \tag{35}$$

Similarly to the flow velocity, other parameters characterizing analyte transport are also constant in the z-direction. This allows us to reduce the transport problem to two spatial dimensions described by the coordinates x and y.

The actual analyte concentration, which is of course no longer constant, is then described as a function of these two coordinates, $\alpha = \alpha(x, y, t)$. The time dependence of the analyte concentration is given by the continuity equation. If no transport mechanism other than the laminar flow is considered, a partial differential equation is obtained:

$$\frac{\partial \alpha(x,y,t)}{\partial t} = -v(y)\frac{\partial \alpha(x,y,t)}{\partial x} = -4v_{\max}\frac{y}{h}\left(1-\frac{y}{h}\right)\frac{\partial \alpha(x,y,t)}{\partial x}\,. \tag{36}$$

To model the effect of an analyte injection, the equation has to be solved for an initial condition of zero analyte concentration inside the flow cell at $t = 0$ and a boundary condition of analyte concentration α_0 at the entrance of the flow cell $\alpha(0, y, t) = \alpha_0$. Results are shown in Fig. 7. It can be seen that for the central part of the vertical profile the injected analyte concentration α_0 is achieved relatively rapidly (depending on the flow rate), but the analyte concentrations remains zero in close proximity of the cell walls. This is a direct consequence of laminar flow – analyte transport to the active surface layer by laminar flow alone is very ineffective.

The other transport mechanism, i.e., translational diffusion of the analyte, becomes therefore highly important in the vicinity of the active sensor layer. Translational diffusion is a mechanism that leads to concentration uniformity in non-mixed solutions. It is described by the first Fick's law that states proportionality between the rate of diffusion and the concentration gradient.

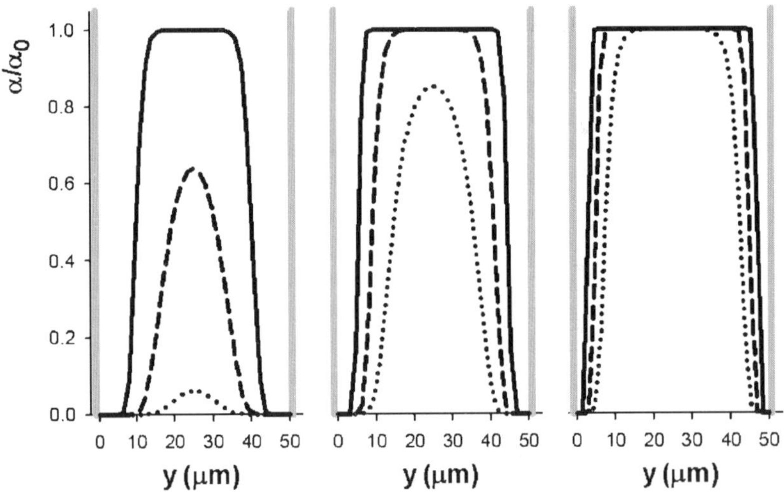

Fig. 7 Analyte concentration in a flow cell at 10 mm distance from the injection entrance reached 3 s (*dotted line*), 6 s (*dashed*), and 15 s (*solid*) after beginning of the injection as a consequence of pure laminar flow (diffusion not considered). Flow rates were 10 μL min^{-1} (*left*), 30 μL min^{-1} (*middle*), and 90 μL min^{-1} (*right*). Cell dimensions: 20 mm (length) × 2.7 mm (width) × 0.05 mm (height)

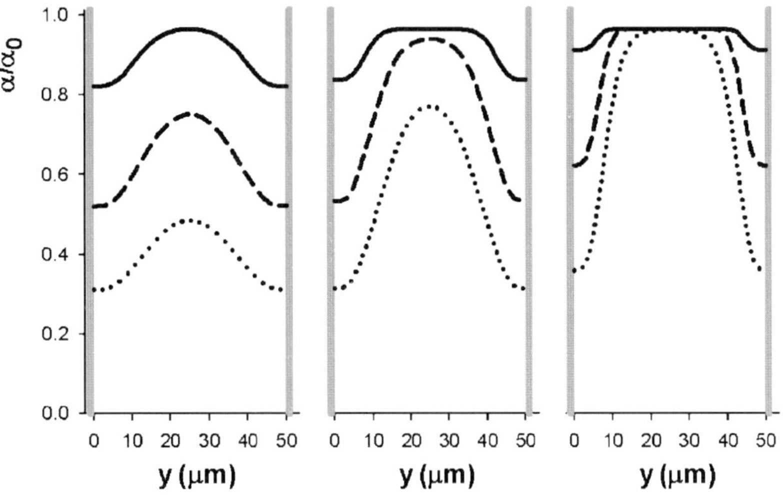

Fig. 8 Analyte concentration in a flow cell at 10 mm distance from the injection entrance reached 3 s (*dotted line*), 6 s (*dashed*), and 15 s (*solid*) after beginning of the injection as a consequence of laminar flow and diffusion. Flow rates were 10 μL min^{-1} (*left*), 30 μL min^{-1} (*middle*), and 90 μL min^{-1} (*right*), diffusion coefficient 10^{-6} cm^2 s^{-1}. Cell dimensions: 20 mm (length) × 2.7 mm (width) × 0.05 mm (height)

The proportionality constant D is called the diffusion coefficient and quantifies the chaotic translation motion of the molecules in solution. Its basic evaluation is given by the Stokes formula (Eq. 33). The diffusion coefficient decreases as the size of the molecule increases. For typical biomolecules in aqueous medium, D is usually between 10^{-7} cm^2 s^{-1} and 10^{-6} cm^2 s^{-1}. Temperature dependence of the diffusion coefficient follows T/η, where T is absolute temperature and η viscosity of the solvent, unless the temperature change does not alter the molecular shape.

If the translation diffusion is taken into account, the equation of the analyte transport will become:

$$\frac{\partial\alpha(x,y,t)}{\partial t} = D\left(\frac{\partial^2\alpha(x,y,t)}{\partial x^2} + \frac{\partial^2\alpha(x,y,t)}{\partial y^2}\right) - 4v_{\max}\frac{y}{h}\left(1-\frac{y}{h}\right)\frac{\partial\alpha(x,y,t)}{\partial x}\,. \quad (37)$$

The effect of diffusion on the analyte distribution is shown in Fig. 8, where Eq. 37 was solved using the same boundary conditions that were applied to Eq. 35 to generate Fig. 7. Note the significant increase in analyte concentration near the cell walls, thanks to diffusion.

3.2 Full Model of Mass Transport

A rigorous approach to modeling the reaction kinetics at the sensor surface, including the mass transport effects, requires solving the fundamental par-

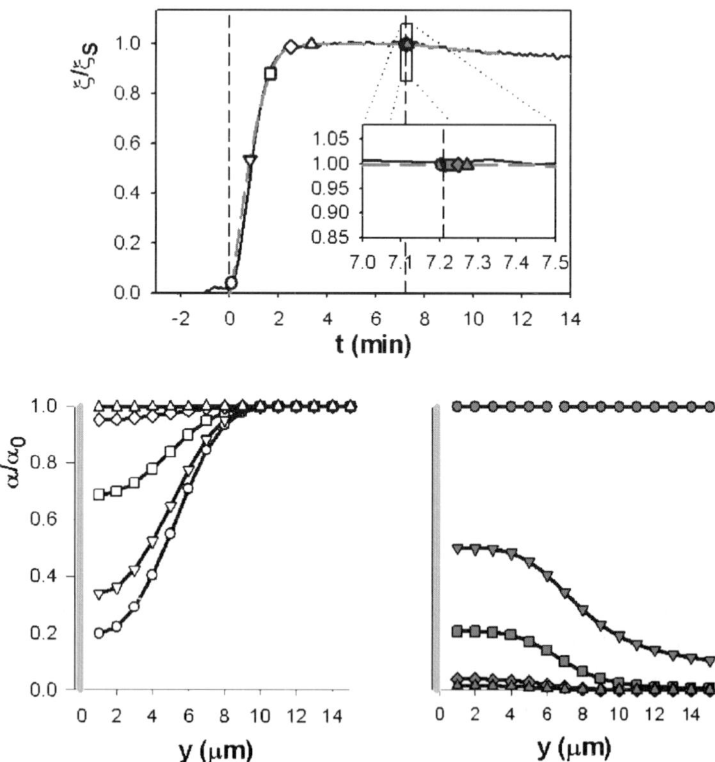

Fig. 9 Results of the full model of the analyte transport coupled to pseudo first-order reaction kinetics. The model fitted to experimental sensorgram of DNA 23-mer binding to its immobilized complementary DNA chain. The experimental sensorgram (*upper graph, black solid line* with the association and dissociation periods indicated by *vertical dashed lines*) is very well fitted by the theoretical course of the relative sensor response (*gray long dashes*). The *graphs below* show the free analyte concentration in the close vicinity of the active sensor layer as it is distributed 6, 50, 100, 150, and 200 s after the injection (on the *left*, from the *bottom up*) and 0, 0.5, 1.12, 2.4, and 3.75 s after stopping the injection (on the *right*, from the *top down*). The times corresponding to particular concentration profiles are indicated in the *upper graph* by same *graphical symbols*. In the case of the dissociation phase they can be resolved only after expansion of the time axis (*insert*). Parameters: $\alpha_0 = 10^{-7}$ M, $\beta = 1.84 \times 10^{-9}$ M cm, $\Phi = 70\,\mu\text{L min}^{-1}$, $D = 2.5 \times 10^{-6}\,\text{cm}^2\,\text{s}^{-1}$, $k_a = 5.6 \times 10^5\,\text{M}^{-1}\,\text{s}^{-1}$, $k_d = 2.5 \times 10^{-6}\,\text{s}^{-1}$. The model was applied to the central part of the flow cell (*xy* coordinates corresponding to the active zone of the sensor) with dimensions of 2.5 mm × 2.7 mm × 0.04 mm

tial differential (Eq. 37, PDE) coupled with the relevant kinetic equations. The coupling is twofold. First, we have to apply the actual analyte concentration at the given point on the sensor surface. As this value varies along the x-coordinate, the concentration of analyte/receptor complexes can no longer be considered as only time dependent – it must be described as a function of t and x: $\gamma = \gamma(x, t)$. The kinetic equations need to be modified accordingly; for instance, the equation of simple first-order kinetics (Eq. 12) is modified as follows:

$$\frac{\mathrm{d}\gamma(x,t)}{\mathrm{d}t} = k_\mathrm{a}\alpha(x,0,t)\,[\beta - \gamma(x,t)] - k_\mathrm{d}\gamma(x,t)\,. \tag{38}$$

Secondly, we have to introduce the consumption or production of the free analyte due to the interaction with receptors to the PDE. It is usually performed [18–20] via a specific boundary condition that in the case of single reaction kinetics is:

$$D\frac{\partial\alpha(x,0,t)}{\partial y} = \frac{\partial\gamma(x,t)}{\partial t}\,. \tag{39}$$

For more complex reaction kinetics the right side of the equation must comprise all kinds of complexes (the formation of which requires consumption of the analyte) multiplied by respective stoichiometric factors. Analogously to Eq. 39, a boundary condition of:

$$\frac{\partial\alpha(x,h,t)}{\partial y} = 0 \tag{40}$$

is introduced for the flow-cell wall opposite to the sensor surface.

Equation 37 can be solved only numerically. Most often a finite element method with various grids in the xy region of the flow cell is employed [18], but other approaches have also been tested [19]. An illustration of the full model results is given in Fig. 9.

The enormously time-consuming nature of full model calculations prevents this approach from being used for complete fits of experimental data. As a rule, it is employed to verify simpler models and/or to confirm the reasonability of rate constants by comparison with experimental data. To enable more convenient and routine analysis of measured sensorgrams, simpler models of mass transport effects have been derived.

3.3 Simplified Models of Mass Transport

The first simplification of Eq. 37 is based on the assumption that the analyte transport in the x direction is mainly conductive, i.e., it is controlled by the flow in the cell. The relation between conductive transport and diffusion in y direction is often characterized by the Peclét number, which reflects the ratio of the ideal time required for an analyte molecule to diffuse from the cell middle ($y = h/2$) to the cell wall, to the minimal time required for that same molecule to pass through the cell by the laminar flow:

$$\mathrm{Pe} = \frac{v_{\max} h^2}{Dl} \,. \tag{41}$$

If the Peclét number is high compared to 1 ($\mathrm{Pe} \gg 1$), Eq. 37 can be simplified by omitting the diffusion term in the x direction and by linearizing the flow velocity dependence on y, because we can limit calculations of the analyte concentration to a region close to the sensor surface ($y \ll h$) [20–22]. We find:

$$\frac{\partial \alpha(x,y,t)}{\partial t} = D \frac{\partial^2 \alpha(x,y,t)}{\partial y^2} - 4 v_{\max} \frac{y}{h} \frac{\partial \alpha(x,y,t)}{\partial x} \,. \tag{42}$$

For a pseudo first-order analyte-to-receptor reaction this equation coupled with the reaction kinetics Eq. 38 via Eq. 39 can be solved so that an equation for only $\gamma(x,t)$ is obtained:

$$\begin{aligned} \frac{\partial \gamma(x,t)}{\partial t} &= k_{\mathrm{a}} \alpha_0 \left[\beta - \gamma(x,t)\right] \\ &\quad \times \left[1 - \frac{Fh}{\alpha_0 D l \mathrm{Pe}^{1/3}} \int\limits_0^x \frac{\partial \gamma(u,t)}{\partial t} (x-u)^{2/3} \mathrm{d}u \right] - k_{\mathrm{d}} \gamma(x,t) \end{aligned} \tag{43}$$

$$F = \frac{1}{12^{1/3} \Gamma(2/3)} \approx 0.32256 \,.$$

Equation 43, which is much easier to solve than the full model, can be further simplified in order to eliminate the integral term. This approximation can be applied when the dependence of $\frac{\partial \gamma(x,t)}{\partial t}$ on x is rather weak and can be assumed to be linear. This linearization allows the integral term to be evaluated explicitly. The result can be written formally in a form analogous to the original kinetic equation (Eq. 12):

$$\frac{\mathrm{d}\gamma(x,t)}{\mathrm{d}t} = k_{\mathrm{a}}^{\mathrm{ef}}(x,t) \alpha_0 \left[\beta - \gamma(x,t)\right] - k_{\mathrm{d}}^{\mathrm{ef}}(x,t) \gamma(x,t) \,, \tag{44}$$

where the "effective" rate constants, however, are both space- and time-dependent:

$$k_{\mathrm{a}}^{\mathrm{ef}} = \frac{k_{\mathrm{a}}}{1 + k_{\mathrm{a}} \left[\beta - \gamma(x,t)\right] / k_{\mathrm{M}}(x)} \qquad k_{\mathrm{d}}^{\mathrm{ef}} = \frac{k_{\mathrm{d}}}{1 + k_{\mathrm{a}} \left[\beta - \gamma(x,t)\right] / k_{\mathrm{M}}(x)} \tag{45}$$

$$k_{\mathrm{M}}(x) \approx 1.034 \left(\frac{v_{\max} D^2}{hx} \right)^{1/3} \,.$$

Thanks to the previously applied assumption that $\gamma(x,t)$ is linearly dependent on x, it is also possible to integrate it over the active sensor region. As a result, we obtain equations analogous to Eq. 44, where $\gamma(x,t)$ is replaced by the average concentration of complexes $\langle\gamma\rangle(t)$:

$$\frac{\mathrm{d}\langle\gamma\rangle(t)}{\mathrm{d}t} = k_{\mathrm{a}}^{\mathrm{ef}}(t)\alpha_0\left[\beta - \langle\gamma\rangle(t)\right] - k_{\mathrm{d}}^{\mathrm{ef}}(t)\langle\gamma\rangle(t)$$

$$k_{\mathrm{a}}^{\mathrm{ef}} = \frac{k_{\mathrm{a}}}{1 + k_{\mathrm{a}}\left[\beta - \langle\gamma\rangle(t)\right]/k_{\mathrm{M}}} \quad k_{\mathrm{d}}^{\mathrm{ef}} = \frac{k_{\mathrm{d}}}{1 + k_{\mathrm{a}}\left[\beta - \langle\gamma\rangle(t)\right]/k_{\mathrm{M}}} \tag{46}$$

$$k_{\mathrm{M}} \approx 1.378\left(\frac{v_{\max}D^2}{hl}\right)^{1/3} .$$

Equations 46 have been directly derived from the full model in [19]. On the other hand, they are almost identical with the relations obtained from the so-called two-compartment model (the only difference is that the numerical coefficient k_{M} is a little bit lower). The two-compartment model was first developed for sensors with receptors placed on small spheres [23]. In [24–26] it was adapted for the SPR flow cell and in [18] it was approved and verified by comparison of numerical results with those obtained from the full model. The two-compartment model approximates the analyte distribution in the vicinity of the receptors by considering two distinct regions. The first is a thin layer around the active receptor zone of effective thickness h_{layer}, and the second is the remaining volume with the analyte concentration equal to the injected one, i.e., α_0. While the analyte concentration in the bulk is constant (within a given compartment), analyte transport to the inner compartment is controlled by diffusion. The actual analyte concentration at the sensor surface is then given by the difference between the diffusion flow and the consumption/production of the analyte via interaction with receptors. For the simple pseudo first-order interaction model we obtain:

$$\frac{\mathrm{d}\alpha}{\mathrm{d}t} = \frac{1}{h_{\mathrm{layer}}}\left[k_{\mathrm{M}}\left(\alpha_0 - \alpha\right) - \frac{\mathrm{d}\gamma}{\mathrm{d}t}\right] . \tag{47}$$

The constant k_{M} can be approximated as [22, 27]:

$$k_{\mathrm{M}} \approx 1.282\left(\frac{v_{\max}D^2}{hl}\right)^{1/3} . \tag{48}$$

For a quasi-steady-state approximation where $\frac{\mathrm{d}\alpha}{\mathrm{d}t}$ is set to zero in Eq. 47, equations analogous to Eq. 46 are obtained from Eqs. 12 and 47.

The k_{M} value can be considered as a measure of the mass transport. Its effect on the SPR response can be evaluated by the maximal difference of the denominator in Eq. 46 from unity [16]. It is equal to the ratio of the reaction velocity to the diffusion flux of the analyte at the beginning of the association stage:

$$\mathrm{denom.}_{\max} - 1 = \frac{k_{\mathrm{a}}\beta}{k_{\mathrm{M}}} \approx 0.780 \qquad k_{\mathrm{a}}\beta\left(\frac{v_{\max}D^2}{hl}\right)^{-1/3} . \tag{49}$$

This ratio (excluding the numerical constants) is called Damköhler number (Da):

$$\mathrm{Da} = k_{\mathrm{a}}\beta \left(\frac{v_{\max} D^2}{hl}\right)^{-1/3} = \frac{k_{\mathrm{a}}\beta h}{D\sqrt[3]{\mathrm{Pe}}} \,. \tag{50}$$

For small Damköhler numbers (Da $\ll$ 1) the mass transport is much faster than the surface reaction itself and therefore the mass transport effect may be ignored. On the other hand, if the Damköhler number is high (Da $\gg$ 1) the sensorgram profile is completely controlled by the diffusion mass transfer and is it not possible to determine rate constants of the surface reaction.

All of the transport models presented so far assume that the diffusion mobility of the analyte in the active sensor zone is the same as in the bulk. In case of the sensors using a thick skeleton to fix the receptors, such as a dextran matrix, solgel, or MIPs, it might be useful to take into account varying analyte diffusion mobility inside the active sensor layer. Detail analysis and proposed models can be found in [28].

4 Summary

A constant and homogeneous concentration of free analyte represents the ideal condition for modeling molecular interactions at the surface of an SPR biosensor. In the most frequent case where an analyte binds to an immobilized receptor with 1 : 1 stoichiometry, the interaction follows the pseudo first-order kinetic model. Adequate interaction models can be built up to describe more complex molecular interactions; some of them have been presented and explained above.

The effect of mass transport on molecular binding in the SPR sensor active layer can be evaluated by means of the Damköhler number (Eq. 50). Except for cases of a very low Damköhler number, mass transport has to be regarded in theoretical models by means of the aforementioned equations.

References

1. Ward LD, Winzor DJ (2000) Anal Biochem 285:179
2. de Mol NJ, Plomp E, Fischer MJE, Ruijtenbeek R (2000) Anal Biochem 279:61
3. Fisher RD, Wang B, Alam SL, Higginson DS, Robinson H, Sundquist WI, Hill CP (2003) J Biol Chem 278:28976
4. McDonnell JM (2001) Curr Opin Chem Biol 5:572
5. Oshannessy DJ, Brighamburke M, Soneson KK, Hensley P, Brooks I (1993) Anal Biochem 212:457
6. Rich RL, Myszka DG (2005) J Mol Recog 18:431
7. Rich RL, Myszka DG (2005) J Mol Recog 18:1

8. Morton TA, Myszka DG (1998) Methods Enzymol: Energetics of biological macromolecules. Pt B 295:268
9. Kamimori H, Unabia S, Thomas WG, Aguilar MI (2005) Anal Sci 21:171
10. Shen BJ, Shimmon S, Smith MM, Ghosh P (2003) J Pharm Biomed Anal 31:83
11. Karlsson R, Falt A (1997) J Immunol Methods 200:121
12. Muller KM, Arndt KM, Pluckthun A (1998) Anal Biochem 261:149
13. Yoo SH, Lewis MS (1995) Biochemistry 34:632
14. Day YSN, Baird CL, Rich RL, Myszka DG (2002) Protein Sci 11:1017
15. Bird RB, Stewart WE, Lightfoot EE (2002) Transport phenomena. Wiley, New York
16. Edwards DA (2000) Stud Appl Math 105:1
17. Brody JP, Yager P, Goldstein RE, Austin RH (1996) Biophys J 71:3430
18. Myszka DG, He X, Dembo M, Morton TA, Goldstein B (1998) Biophys J 75:583
19. Mason T, Pineda AR, Wofsy C, Goldstein B (1999) Math Biosci 159:123
20. Edwards DA (1999) IMA J Appl Math 63:89
21. Edwards DA, Goldstein B, Cohen DS (1999) J Math Biol 39:533
22. Lok BK, Cheng YL, Robertson CR (1983) J Colloid Interface Sci 91:104
23. Glaser RW (1993) Anal Biochem 213:152
24. Myszka DG, Morton TA, Doyle ML, Chaiken IM (1997) Biophys Chem 64:127
25. Schuck P, Minton AP (1996) Anal Biochem 240:262
26. Schuck P (1996) Biophys J 70:1230
27. Sjolander S, Urbaniczky C (1991) Anal Chem 63:2338
28. Edwards DA (2001) Bull Math Biol 63:301

Part II
Implementations of SPR Biosensors

Springer Ser Chem Sens Biosens (2006) 4: 95–116
DOI 10.1007/5346_016

Published online: 8 July 2006

SPR Sensor Instrumentation

Marek Piliarik · Jiří Homola (✉)

Institute of Radio Engineering and Electronics, Prague, Czech Republic
homola@ure.cas.cz

Keywords Data processing · Multichannel sensors · Noise analysis · Optical sensor · Sensor instrumentation · Spectroscopy of surface plasmons · SPR imaging · Surface plasmon resonance

1 Introduction

The potential of surface plasmons for optical sensing was recognized in the early 1980s when surface plasmons, excited in the Kretschmann geometry of the attenuated total reflection method, were used to probe processes at the surfaces of metals [1] and to detect gases [2]. Since then, numerous surface plasmon resonance (SPR) sensors have been reported.

An SPR sensor instrument consists of an optical system, supporting electronics, and a sensor data acquisition and processing system. In the optical system, surface plasmons are optically excited and the output light wave with an encoded SPR signal is detected. The signal from the detector is processed to yield a sensor output. SPR biosensors also incorporate a biorecognition

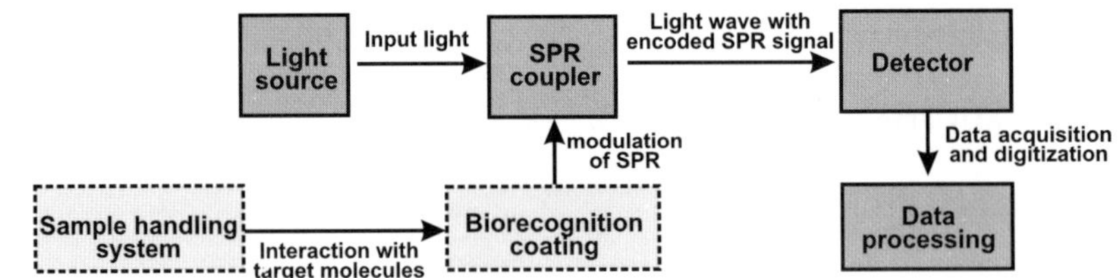

Fig. 1 Scheme of an SPR (bio)sensor

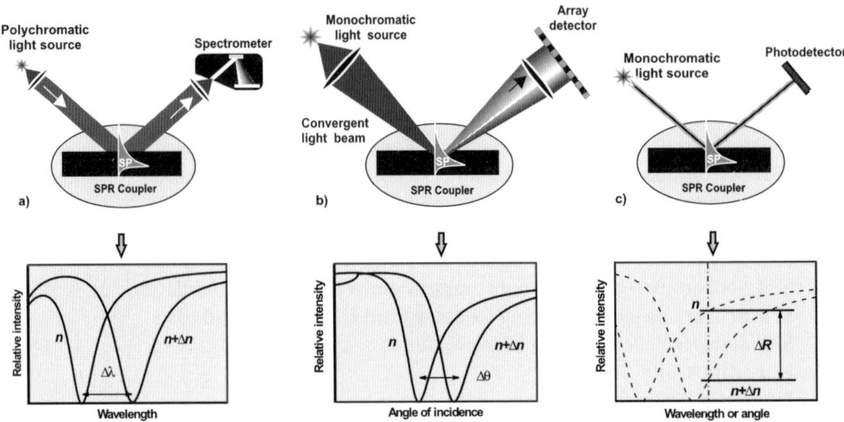

Fig. 2 SPR sensors based on modulation of **a** coupling angle, **b** coupling wavelength, and **c** light intensity

coating that interacts with target molecules in a liquid sample, and a sample preparation and handling system (Fig. 1).

In the optical system of an SPR sensor, surface plasmons are excited by a light wave. The excitation of surface plasmons in the SPR sensor results in a change in one of the characteristics of the light wave. Based on which characteristic of the light wave is modulated and used as a sensor output, SPR sensors can be classified as SPR with (i) angular, (ii) wavelength, (iii) intensity, (iv) phase, or (v) polarization modulation. The first three types of modulation (Fig. 2) are used most frequently in today's SPR sensors.

2 Data Processing for SPR Sensors

In wavelength (or angular) modulation-based SPR sensors, light emerging from the SPR coupler is dispersed over a detector array, illuminating each pixel with light of a slightly different wavelength (or associated with a slightly different angle of incidence in the SPR coupler). The detector is digitized periodically, producing a series of spectra in time. The sets of intensity values are mathematically processed to determine the sensor output. In the first phase of data processing, the spectra are usually averaged and normalized. The averaging either involves averaging of time series of intensity from the same detector (time averaging) or averaging of intensities from multiple detectors (e.g., of a 2D array) measured at a single time (spatial averaging). In wavelength or angular modulation-based SPR sensors, it is usually applied to spectra that are measured in several rows of a 2D array detector [3, 4]. The spectra contain SPR information in a rather raw form, as the shape of the spectra are influenced by a variety of other effects (e.g., properties of emission spectrum of the light source and SPR optics). In order to reconstruct the SPR spectrum, the spectra are normalized. First, the light intensity is corrected by subtracting the dark signal (intensity measured in the absence of light) caused by leakage current in the detector. Then, the resulting signal is divided by a reference signal to compensate for uneven (angular or spectral) distribution of illumination and absorption in the optical system [5–7]. As surface plasmons are TM-polarized waves, it is common to use a TE-polarized light for the reference, although other non-resonant signals may be also used. An example of raw spectra acquired from an SPR sensor with wavelength modulation and the resulting normalized spectrum are given in Fig. 3.

Subsequently, a metric which can be associated with the angular or wavelength position of an SPR dip is calculated from angular or wavelength normalized spectrum. Numerous algorithms have been developed to calculate the sensor output. One of the most commonly utilized algorithms is the centroid method, which calculates the centre-of-mass of a portion of SPR dip that

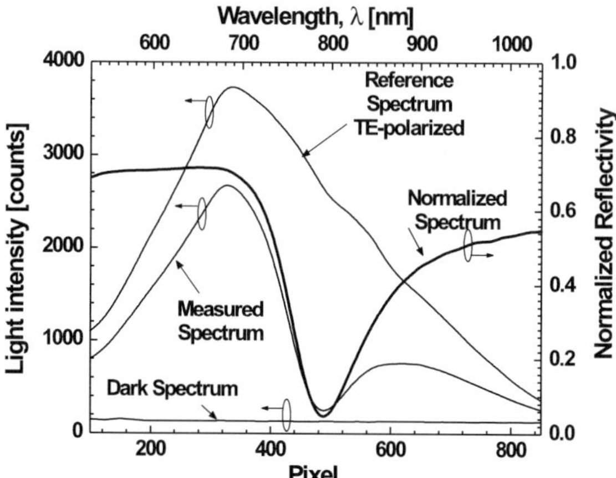

Fig. 3 Raw measured and reference spectra and a normalized spectrum with an SPR dip for a wavelength modulation-based sensor using a linear detector array [6]

falls below a certain threshold (set typically around half of the depth) [8, 9]. The issue of proper accounting of intensity values from pixels that enter or leave the range below threshold was addressed by the weighted centroid [7] or partial pixel accounting [6] approaches. Modifications of the centroid method improving resistance of the algorithm to overall light level fluctuations were also proposed [6, 10]. Another approach to calculating the SPR dip position consists of fitting a polynomial to a certain portion of the SPR dip and subsequently calculating the minimum of the polynomial [11, 12]. In terms of sensor output noise, this algorithm performs comparably to the centroid method if the same region of the SPR dip is considered. Chinowsky et al. proposed the optimal linear data analysis method which determines the sensor output by an optimized linear transformation of the difference of the reference and measured spectrum [13]. This method provided a sensor output with noise which was smaller than that generated using the centroid method by a factor of 1.3. Recently, the model parametrization and linear projection (MPLP) method was introduced [14]. In this method, the spectrum shape is described by a set of parameters that are then algebraically manipulated directly into the measurand. Experimental investigation of MPLP method performance yielded a decrease of the sensor output noise by a factor of 1.7 when compared to the centroid method, and 1.3 compared to the polynomial fitting method [14].

In SPR sensors with intensity modulation, the intensity value or its distribution is detected by an individual detector or a matrix of detectors. The resulting set of intensities is usually averaged in time and space [15–17], leading directly to the sensor output.

3 Optical Systems for SPR Sensors

In the optical system of an SPR sensor, surface plasmons are optically excited and the SPR signal is encoded into a light wave interacting with the surface plasmons. Based on the method of excitation of surface plasmons, SPR sensors can be classified as SPR sensors based on (i) prism couplers, (ii) grating couplers, and (iii) waveguide couplers (Fig. 4).

3.1 SPR Sensors Based on Prism Couplers

Most of the SPR sensors reported to date are based on prism couplers. Prism coupling of light into surface plasmons is convenient and can be realized with simple and conventional optical elements. It can be readily combined with any type of modulation. Specific examples of prism-based SPR sensors are discussed below.

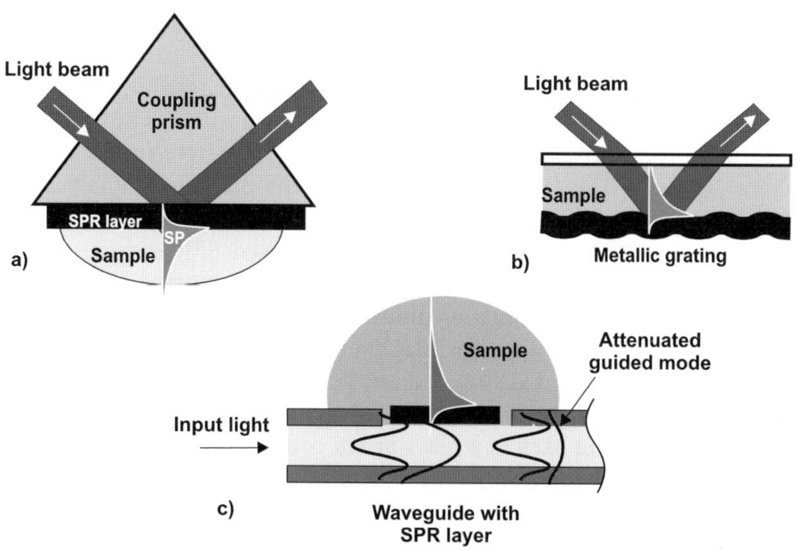

Fig. 4 SPR sensors based on **a** prism couplers, **b** grating couplers, and **c** waveguide couplers

3.1.1
SPR Sensors Based on Prism Couplers and Angular Modulation

In 1988 Matsubara et al. reported an SPR sensor based on prism coupler and angular modulation [18]. The optical system of their sensor is shown in Fig. 5.

Matsubara et al. demonstrated that the use of a lens and photodiode array placed in the back-focal plane of the lens makes it possible to reconstruct

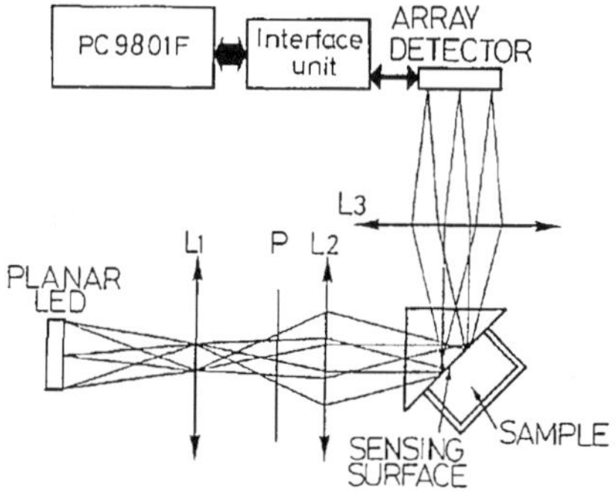

Fig. 5 Schematic diagram of SPR sensor with angular modulation. Lenses *L1* and *L2* provide an angular span of incident light and lens *L3* projects the angular spectrum in the back focal plane on the detector array. Reprinted from [18], copyright 1988, with permission from Optical Society of America

the angular spectrum of the reflected divergent beam. This design allows the use of a cheap surface emitting LED as a light source without the degradation of the angular resolution. In this work, an angular resolution of 0.01 deg was achieved. This corresponds to a bulk refractive index resolution of about 5×10^{-5} RIU (refractive index unit) assuming a refractive index sensitivity of about 200 deg RIU^{-1} (this value is based on the theoretical analysis of the investigated sensor design).

In the early 1990s, an angular modulation-based SPR sensor with a refractive index resolution of about 2×10^{-6} RIU was reported [5, 11, 19]. The sensor consisted of a light-emitting diode (LED, wavelength – 760 nm), a glass prism and a detector array with imaging optics (Fig. 6). A divergent beam produced by the LED was collimated and focused by means of a cylindrical lens to produce a wedge-shaped beam of light that was used to illuminate a thin gold film on the back of a glass prism containing several sensing areas (channels). The imaging optics consisted of one imaging and one cylindrical lens ordered in such a way that the angular spectrum of each sensor channel was projected on a separate row (or rows) of the array detector.

This optical design has been further advanced by Biacore (Pharmacia Biosensors AB; since 1996, Biacore AB) and resulted in a family of commercial SPR sensors [20–22] offering high performance (resolution down to 1×10^{-7} RIU) and multiple sensing channels (up to four) for simultaneous measurements.

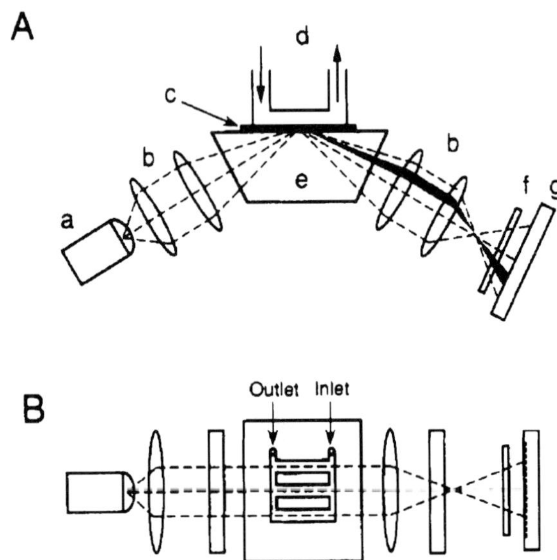

Fig. 6 SPR sensor in angular configuration with three parallel channels (**A** *side view*, **B** *top view*). *a* High output light emitting diode, *b* lenses, *c* sensor chip, *d* microfluidic cartridge, *e* coupling prism, *f* polarizer, *g* photodiode array detector. Reprinted from [11], copyright 1991, with permission from American Chemical Society

In 2002, Zhang et al. reported a simple, effective and self-referenced configuration of an SPR with angular modulation that utilized a quadrant cell photodetector instead of a photodiode array [23, 24]. A linearly polarized laser beam (wavelength 635 nm) was focused in one direction and directed onto a thin gold film containing sensing and reference measuring areas (Fig. 7). The reflected divergent beam contained the whole range of angles of incidence. A four-cell photodetector was placed into the reflected light in such a way that the sensing channel illuminated one pair of cells and the reference channel another pair of cells. The pair of cells divided the resonant dip approximately in the minimum of reflectivity and only four intensities were measured and processed at a time. When the angular position of the dip changed, the ratio of the intensities changed. The use of a detector with large area photodiode cells made it possible to achieve extremely low noise levels. In addition, as the two cells were nearly identical, certain common noises (e.g., noise caused by laser intensity fluctuations) were compensated for by this approach leading to a demonstrated angular resolution as low as 10^{-5} deg. Considering the sensor sensitivity of about 130 deg RIU^{-1}, this translates to a refractive index resolution of about 10^{-7} RIU [23]. This high resolution can, however, be delivered within a rather limited operating range, typically about 10% of the width of an SPR dip, which corresponds to a refractive index range of about 4×10^{-3} RIU.

Thirstrup et al. demonstrated an integration of several optical elements into a single sensor chip [25]. In this approach, the cylindrical focusing optics utilized to create a beam of a desired angular span is replaced by a diffraction grating of a special design incorporated into the sensing element Fig. 8 [26]. A wide parallel light beam (wavelength 670 nm) was diffracted by the focusing (chirped) grating and focused into small spot on the SPR measuring surface. The reflected light followed a similar path, producing a parallel beam with an angular spectrum superimposed across the beam. A 2D CMOS detec-

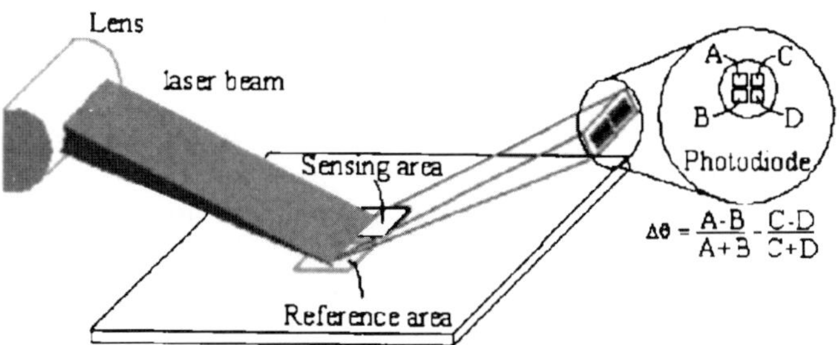

Fig. 7 Differential SPR sensor. A quadrant cell photodetector simultaneously measures the SPR dips from the reference and sample areas, and the difference signal provides an accurate sensor output. Reprinted from [23], copyright 2003, with permission from American Institute of Physics

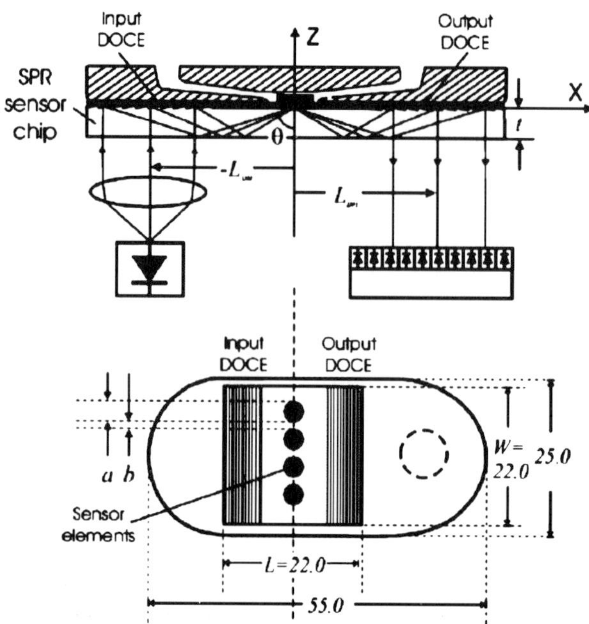

Fig. 8 An SPR coupling chip with integrated diffractive elements replacing the focusing and imaging optics. Reprinted from [25], copyright 2004, with permission from Elsevier

tor measured the angular spectrum as a distribution of light intensity along a row of detector pixels for several parallel channels (using different columns of pixels of the detector). This design offered a compact SPR platform with a refractive index sensitivity of 130 deg RIU^{-1} and a refractive index resolution of about 5×10^{-7} RIU.

A miniature version of the angular modulation-based SPR sensor was reported by Melendez et al. who integrated all electro-optical components of the SPR sensor into a small monolithic platform [27]. Figure 9 depicts an advanced version of this SPR sensor design (Spreeta 2000 by Texas Instruments). The sensor consisted of a plastic prism molded onto a microelectronic platform containing an infrared LED (830 nm peak wavelength) and a 128-pixel linear diode array detector. The LED emitted a diverging beam that passed through a polarizer and struck the sensor surface at a range of angles. The angle at which light reflected from this surface varied with the location on the surface. The light reflected from the sensor's top mirror and back down onto the diode array. The diode array measured the angular spectrum of reflected light.

Baseline noise and smoothness of response of this sensor were investigated by Chinowsky et al. [28]. They concluded that the baseline noise established under constant conditions was less than 2×10^{-7} RIU; however, the sensor response to a gradual change in the refractive index revealed departures from

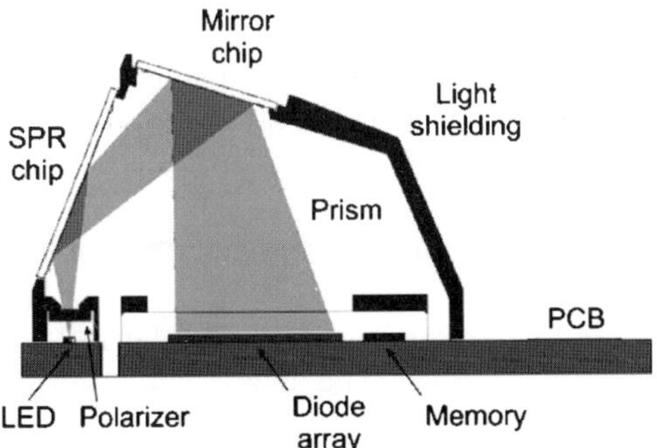

Fig. 9 Cross section of the Spreeta 2000, showing components of the sensor and the path followed by light inside the sensor. The dimensions of the sensor are 1.5 cm × 0.7 cm × 3 cm. Reprinted from [28], copyright 2003, with permission from Elsevier

the expected sensor output of about 0.2% for a refractive index interval of 0.04 RIU (which corresponds to 8×10^{-5} RIU).

Angular spectroscopy of surface plasmons was introduced into a spatially resolved measurements by Kano et al., who demonstrated SPR scanning microscopy in 1998 [29, 30]. Surface plasmons were excited with a high numerical aperture microscope objective that focused the illuminating light via immersion oil and substrate on a silver film (Fig. 10). Part of the light hitting the metal film at a coupling angle coupled to a surface plasmon on the thin film, while the rest of the light was reflected back. The reflected light was collected with the same objective. The angular dip in the reflectivity was observed as a dark ring in the back focal plane that was projected on a CCD camera. This approach provides a highly localized measurement and,

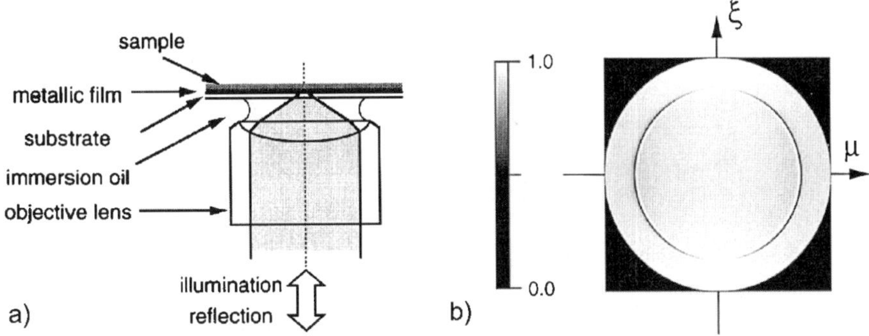

Fig. 10 Optical arrangement of SP excitation in scanning SPR microscope (**a**) and the intensity distribution at the exit pupil (frequency domain) (**b**). Reprinted from [32], copyright 2004, with permission from Elsevier

when combined with a scanning mechanism, SPR sensing with a high spatial resolution [31]. Using this approach, Kano and Knoll were able to observe particles as small as 1.5 μm in diameter [32].

3.1.2
SPR Sensors Based on Prism Couplers and Wavelength Modulation

A convenient modular SPR sensor based on wavelength modulation was developed by Homola et al. [33–35]. The sensor consisted of a halogen lamp, SPR sensor platform, and spectrometer (Fig. 11). White light from the halogen lamp was brought to the SPR sensor platform via a multimode optical fiber. The sensor platform comprised an input collimator producing a large diameter parallel beam, a glass prism with an attached SPR chip (coated with a 50 nm thick gold layer), polarizer, and multichannel output collimator. The output collimators coupled the light into optical fibers, which were connected to inputs of the spectrograph (Fig. 11).

Further optimization of this approach allowed the sensor to resolve changes in the coupling wavelength as low as 1.5×10^{-3} nm, which for the sensor sensitivity of 7500 nm RIU^{-1}, translates to a refractive index resolution of 2×10^{-7} RIU [6].

An alternative geometry of this design, employing a planar optical lightpipe instead of the prism coupler, has been proposed by Nenninger et al. with the goal of eliminating the need for refractive index matching between the prism coupler and a chip [36]. The optical layout of this sensor is shown in Fig. 12. White light from a polychromatic light source was coupled in and out of the lightpipe by means of optical prisms. When propagating in the

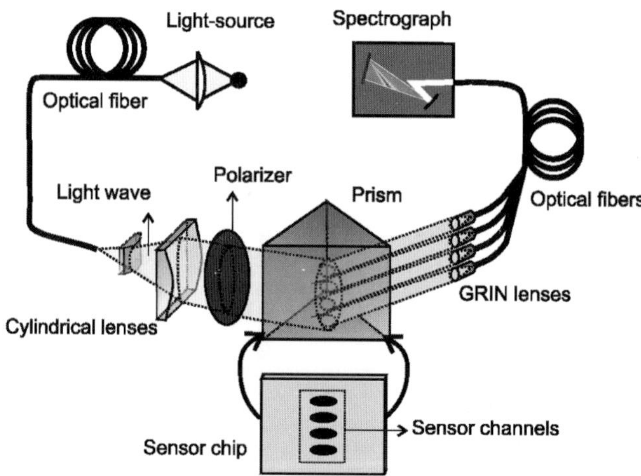

Fig. 11 SPR sensor with wavelength modulation in four-channel configuration [75]

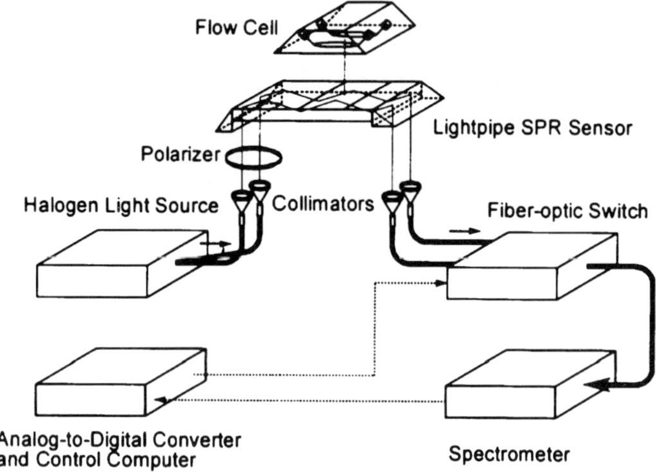

Fig. 12 A dual-channel lightpipe SPR sensor with wavelength modulation. Reprinted from [36], copyright 1998, with permission from Elsevier

lightpipe, the light excited surface plasmons in a central portion of the lightpipe coated with a thin gold film. The light transmitted through the lightpipe was detected and analyzed by a spectrograph. At the operating wavelength of 630 nm, a refractive index resolution of 6×10^{-6} RIU was achieved [36].

In order to increase the amount of information in SPR sensors with wavelength modulation, the wavelength division multiplexing (WDM) approach was proposed by Homola et al. [37]. In this approach, signals from multiple surface plasmons excited in different areas of a sensing surface are encoded into different regions of the spectrum of the light wave. Two configurations of WDMSPR sensors have been reported [38, 39]. In the first approach, a wide parallel beam of polychromatic light is made incident onto a sensing surface consisting of a thin gold film, a part of which is coated with a thin dielectric film (tantalum pentoxide) (Fig. 13a). As the presence of the thin dielectric film shifts the coupling wavelength to a longer wavelength (compared to the bare gold) the reflected light exhibits two dips (Fig. 13c) associated with the excitation of surface plasmons in the area with and without the overlayer [38]. The second approach to WDMSPR sensing uses a special kind of prism in which a polychromatic light is sequentially incident on different areas of the sensing surface at different angles of incidence (Fig. 13b) [39]. Due to the different angles of incidence, the surface plasmons in each region are excited with a different wavelength of the incident light. Therefore, the spectrum of transmitted light contains multiple dips associated with surface plasmons in different areas of the sensing surface (Fig. 13c). A combination of the WDMSPR approach with the conventional parallel architecture, leading to an eight-channel SPR sensor was also demonstrated [39]. The bulk refractive index sensitivities of the two WDM channels

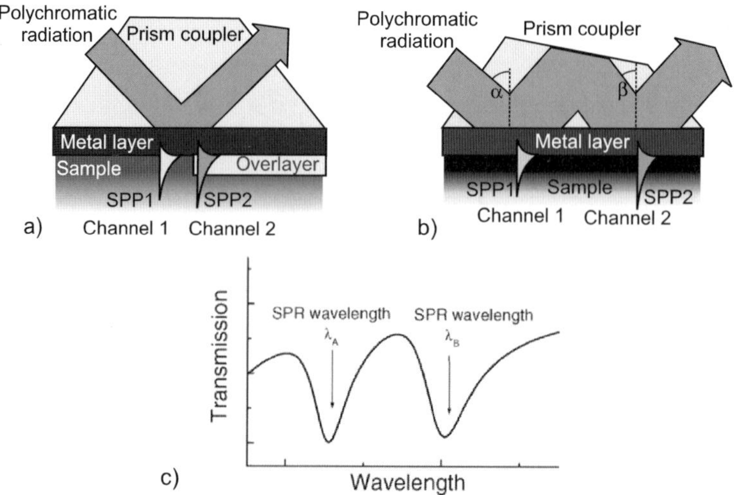

Fig. 13 SPR sensors with wavelength division multiplexing. **a** Spectral encoding by means of a high refractive index overlayer [38]. **b** Spectral encoding by means of altered angles of incidence [39]. **c** Spectral reflectivity with two SPR dips

was $S_B(640\,\text{nm}) = 2710\,\text{nm RIU}^{-1}$, $S_B(790\,\text{nm}) = 8500\,\text{nm RIU}^{-1}$, and refractive index resolutions of 1.3×10^{-6} RIU and 7×10^{-7} RIU, respectively, were achieved [39].

A miniaturized SPR sensor with wavelength modulation and a retro-reflecting prism was proposed by Cahill et al. [40]. Polychromatic light was coupled into a prism of a special shape using an optical fiber and collimating lens (Fig. 14). Reflected light was collected with the same optics to the illuminating fiber and separated from the incident light by means of a fiber coupler. A refractive index sensitivity of about $6000\,\text{nm RIU}^{-1}$ and a refractive index resolution of 3×10^{-5} RIU were demonstrated with this type of SPR sensor. An advanced design of this sensor employing the WDM overlayer approach was demonstrated to provide a better sensor stability [41].

In order to improve the performance of SPR sensors, research into long-range surface plasmons (symmetrical SP) have been pursued. An SPR sensor exploiting long-range surface plasmons was demonstrated by Nenninger et al. [42]. In this work, a long-range surface plasmon was excited on a special multilayer structure consisting of a glass substrate, 700 nm thick Teflon AF layer, and 24 nm thick gold layer. A refractive index sensitivity of $30\,000\,\text{nm RIU}^{-1}$ was reported at a wavelength of 690 nm, resulting in a refractive index resolution of about 2×10^{-7} RIU [42]. Most recently, Slavík and Homola demonstrated a sensor based on an asymmetric multilayer structure employing both the symmetric and antisymmetric surface plasmons supported by a thin gold film [43].

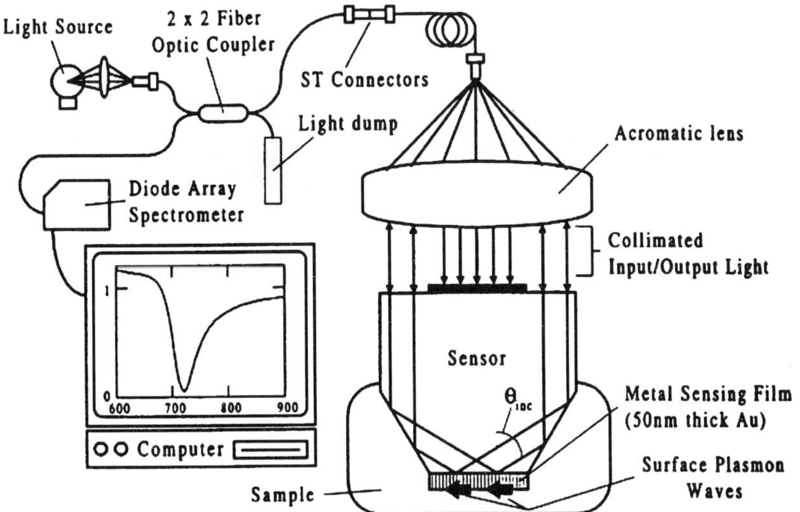

Fig. 14 SPR sensor based on a retro-reflecting probe. Reprinted from [40], copyright 1997, with permission from Elsevier

3.1.3
SPR Sensors Based on Prism Couplers and Intensity Modulation

While the first SPR sensors with prism coupling used intensity modulation [2], since the late 1980s, this modulation approach has become particularly attractive for the development of SPR sensing devices for spatially resolved measurements. The first type of SPR sensor with spatial resolution is an SPR imaging sensor [44, 45]. In SPR imaging, a parallel TM-polarized beam of monochromatic light is launched into a prism coupler and made incident on a thin metal film at an angle of incidence close to the coupling angle for the excitation of surface plasmons. The intensity of reflected light depends on the strength of the coupling between the incident light and the surface plasmon and therefore can be correlated with the distribution of the refractive index along the metal film surface. This approach allows creation of sensing devices with a large number of sensing areas – sensing channels (> 100) [15, 17, 46]. In order to increase the sensor stability and optimize the contrast of SPR images, Fu et al. introduced an SPR imaging setup employing a white light source and a bandpass interference filter [47]. Their SPR sensor instrument, operating at a wavelength of 853 nm, was demonstrated to provide a refractive index resolution of 3×10^{-5} RIU [17]. The spatial resolution of this system was better than 50 μm.

Recently, Piliarik et al. proposed an SPR imaging configuration based on the combination of SPR imaging with polarization contrast and a spatially patterned multilayer SPR structure [15]. In this configuration, a prism coupler with a special patterned multilayer structure was placed between two crossed polarizers (Fig. 15). The output polarizer blocked all the light reflected from the (inactive) areas outside the sensing pads, generating high-contrast images. Two types of SPR pads with opposite sensitivities to refractive index were employed. The reflected light was imaged on a 2D CCD detector and the ratio of the intensities generated from neighboring pads was used to provide a sensor output immune to changes in the intensity of the emitted light. This sensor was demonstrated to provide refractive index resolution better than 5×10^{-6} RIU in more than 100 sensing channels (size of each sensing spot was 400×800 μm).

A dual-wavelength SPR imaging was proposed by Zybin et al. [48], who used two laser diodes which were switched on sequentially, and the intensities of the reflected light at the two different wavelengths were measured (Fig. 16). The sensor output was defined as a difference of these two signals. A refractive index resolution as low as 2×10^{-6} RIU was achieved when averaging over a large beam diameter (6 mm^2) was used.

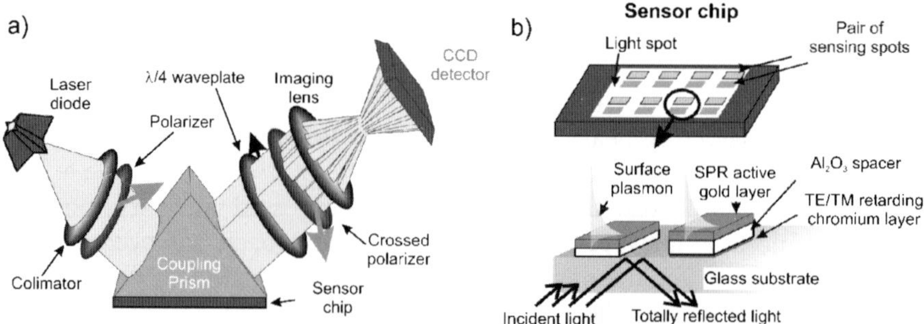

Fig. 15 SPR imaging using polarization contrast and special SPR multilayer structures with crossed sensitivities. The setup layout (**a**) and the sensor chip with measuring pads (**b**) [15]

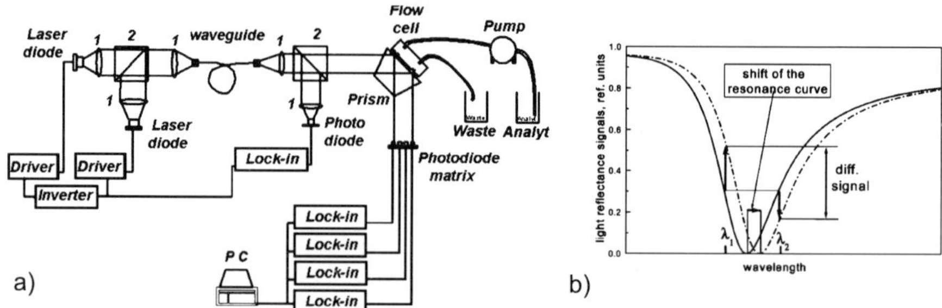

Fig. 16 Dual-wavelength SPR imaging (**a**) and spectrum of the reflected light (**b**). Reprinted from [48], copyright 2005, with permission from American Chemical Society

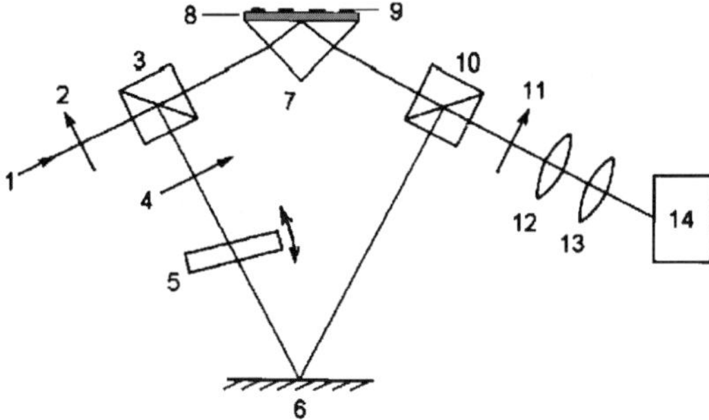

Fig. 17 SPR interferometric sensor: *1* light beam, (*2, 4*) polarizer, (*3, 10*) beam-splitting cubes, (*5*) phase-retarding glass plate, (*6*) mirror, (*7*) SPR prism, (*8*) gold film, (*9*) patterned coating, (*11*) analyzer, (*12, 13*) imaging lenses, (*14*) CCD camera. Reprinted from [49], copyright 2000, with permission from Elsevier

Nikitin et al. developed two interferometric approaches to SPR imaging [49]. The first approach was based on a Mach–Zehnder interferometer combining TM-polarized signal and reference beams (Fig. 17). The second method was based on the interference of the TM-polarized signal beam with the TE-polarized reference beam [50]. This configuration was demonstrated in two modes: (a) phase contrast (Zernike phase contrast) increasing the sensor sensitivity and (b) "fringe mode", in which there was a definite angle between the interfering beams and a pattern of interference fringes was superimposed to the image of the surface. Local variations in the phase of the signal beam reflected from the surface resulted in bending and moving the interference fringes. This approach allowed resolving a refractive index change in the order of 10^{-7} RIU. A similar configuration was used for gas flow imaging by Notcovich et al., who achieved a refractive index resolution of about 10^{-6} RIU [51].

3.2 SPR Sensors Based on Grating Couplers

Although to date grating couplers have been used in SPR to a lesser extent than have prism couplers, they offer some very attractive features. Most importantly, as gratings can be fabricated by techniques such as replication into plastic substrates, grating-based SPR chips provide an avenue to low-cost SPR sensors.

3.2.1 SPR Sensors Based on Grating Couplers and Angular Modulation

Dostalek et al. reported a grating-based SPR sensor with angular modulation and a high number of sensing channels suitable for high-throughput

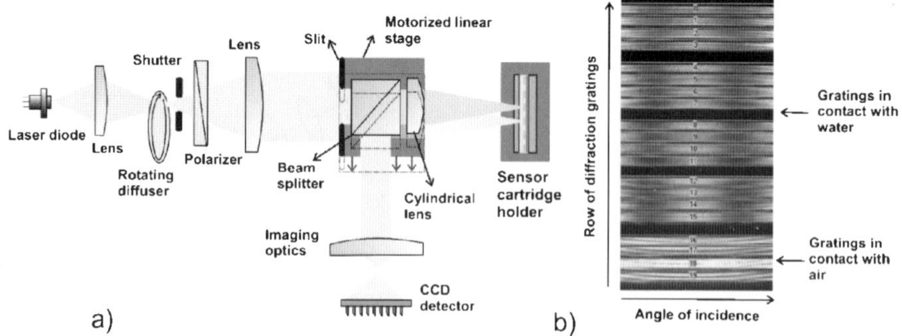

Fig. 18 Grating-based SPR sensor with angular modulation for high throughput screening applications (**a**) and a typical image with two SPR minima (**b**) [52]

screening applications [52]. In this configuration, a collimated beam of monochromatic light was focused with a cylindrical lens on a row of gold-coated diffraction gratings and reflected under nearly normal incidence (Fig. 18). The angular spectra were transformed back to a collimated beam by means of a focusing lens and projected onto a 2D CCD detector. Different rows of gratings were read sequentially by moving the beam splitter and cylindrical lens with respect to the sensor chip. Due to the normal incidence and symmetry of the structure, two SPR dips were observed for each diffraction grating. The sensor was operated at wavelength 635 nm and provided an angular sensitivity of 80 deg RIU^{-1}. A refractive index resolution of 5×10^{-6} RIU was achieved for simultaneous measurements in over 200 sensing channels [52].

3.2.2
SPR Sensors Based on Grating Couplers and Wavelength Modulation

Jory et al. demonstrated an SPR sensor based on a grating coupler and wavelength modulation [53]. A collimated beam of polychromatic light was made incident on a metal-coated grating. The reflected beam was collected and directed to a spectrum analyzer. In order to improve the accuracy of the measurement of the coupling wavelength, an approach combining the wavelength modulation with an acousto-optic tunable filter (AOTF) was introduced (Fig. 19) [54]. The AOFT was utilized to modulate the wavelength of narrow-band incident light around the resonant wavelength ($\sim$ 636 nm). A differential reflectivity profile, correlated with the incident wavelength, was registered by a detector. By locking to the zero differential corresponding to the SPR reflectivity minimum and monitoring the AOTF drive frequency, the SPR minimum position was measured with an accuracy better than 0.0005 nm. This accuracy in determining the

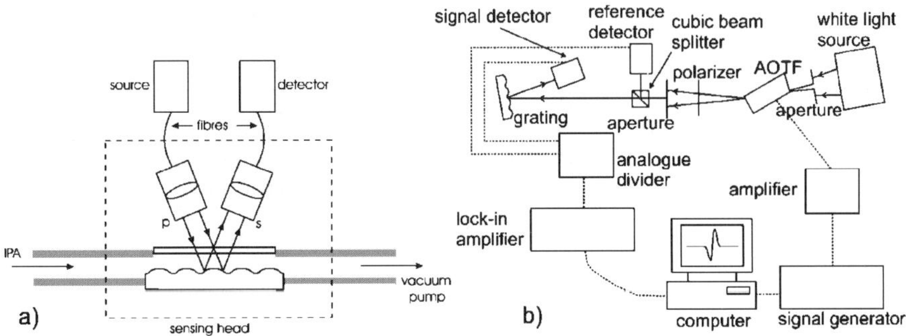

Fig. 19 Grating coupler SPR sensor with wavelength modulation (**a**) and its improvement using AOTF (**b**). Reprinted from [53], copyright 1994, with permission from Elsevier and [54], copyright 1995, with permission from Institute of Physics Publishing

coupling wavelength translates to a refractive index resolution better that 1×10^{-6} RIU.

Recently, an SPR sensor using a wavelength division multiplexing (WDM) on a multidiffractive grating was reported [55]. A polychromatic light beam was made incident onto a special metallic grating with a grating profile composed of multiple harmonics. The reflected light contained multiple SPR dips, one for each grating period. By probing refractive index changes at the sensor surface using multiple surface plasmons of different field profiles, it is possible to distinguish surface refractive index changes (e.g., due to the binding of analyte to the biorecognition elements immobilized on the sensor surface) from background refractive index variations (e.g., due to the sample composition variations). A refractive index resolution of about 5×10^{-6} RIU was achieved with this type of sensor [56].

3.2.3
SPR Sensors Based on Grating Couplers and Intensity Modulation

Brockman et al. presented an SPR imaging device with grating coupled SPR in a microarray of measuring channels. This microarray imaging approach was capable of parallel analysis of spatially distributed information along the sensor surface [57]. A collimated monochromatic light beam (wavelength 860 nm) was made incident onto a plastic chip with a gold-coated diffraction grating. An array of 400 sensing channels was prepared on the chip by means of spatially resolved functionalization (spots 250 μm in diameter). Upon reflection from the chip, the light was projected onto a 2D CCD array (Fig. 20). This approach was commercialized by HTS Biosystems in the flex-chip system [57].

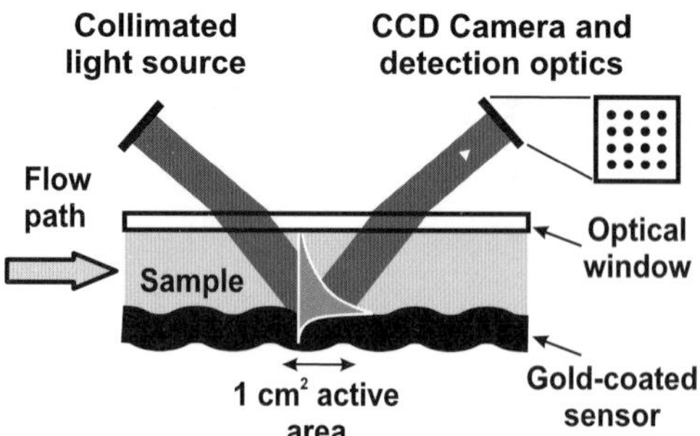

Fig. 20 Concept of an SPR sensor based on a grating coupling and intensity modulation (FLEX CHIP) [57]

3.3
SPR Sensors Based on Optical Waveguides

Over the last 15 years, numerous SPR sensors based on optical fibers or integrated optical waveguides have been proposed. Waveguide-based SPR sensors use either wavelength or intensity modulation and offer compact and miniature sensing elements with the ability to perform localized measurements in hard-to-access locations.

3.3.1
Fiber Optic SPR Sensors

A fiber optic SPR probe with wavelength interrogation was proposed by Jorgenson and Yee [58]. The sensor consisted of a multimode optical fiber with locally exposed core and a thin gold film evaporated around it. A mirror at the end of the sensing area reflected the light back to the fiber and a fiber optic coupler was used to separate the reflected light from incoming illumination [59]. The refractive index resolution reported with this sensor was about 5×10^{-5} RIU. Truilett et al. reported a fiber optic SPR sensor using a similar structure (a multimode optical fiber with locally removed cladding and gold layer deposited around the fiber core). However, the fiber was illuminated by a monochromatic light through special optics to selectively excite fiber modes within a narrow angular span of incident light, and detected changes in the intensity of transmitted light [60]. A refractive index resolution better than 8×10^{-5} RIU was reported with this type of sensor. The major drawback of SPR sensors based on multimode optical fibers is the modal and polarization conversion due to perturbations of the fiber (e.g., bends or defects) which limit the stability of the sensor output. In order to overcome this problem, SPR sensors based on single-mode optical fibers were developed [61, 62]. Single-mode optical fibers support only one mode of the electromagnetic field and therefore no modal conversion can occur. A fiber optic SPR sensor

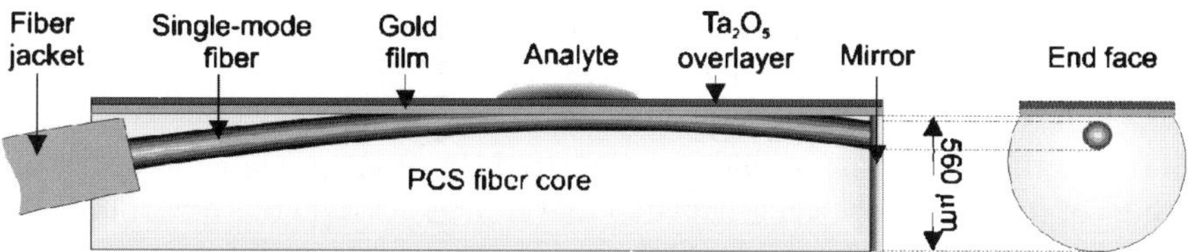

Fig. 21 SPR fiber optic probe using a side-polished single-mode optical fiber [63]

using a side-polished single-mode optical fiber with a thin metal overlayer was developed by Homola [62]. Later, this geometry was reduced to a miniature SPR fiber optic probe (Fig. 21) [63].

An intensity-modulated version of such a sensor was shown to provide a refractive index resolution better than 2×10^{-5} RIU. A wavelength-modulated version of this sensor exhibited even better resolution – 5×10^{-7} RIU [64]. However, these SPR sensors suffered from polarization instability, even when light depolarization was introduced. To eliminate this source of instability, SPR sensors using polarization-maintaining fibers were introduced [65]. These sensors were demonstrated to provide a refractive index resolution of about 2×10^{-6} RIU.

3.3.2 Integrated Optical SPR Sensor

Since the demonstration of the first integrated optical SPR sensor by researchers at the University of Twente in the late 1980s [66], integrated optical SPR sensors using slab waveguides [67] and channel waveguides [68] have been developed.

An integrated optical SPR sensor with intensity modulation and one sensing and one reference channel was reported by Mouvet et al. [68]. The signal from the sensing channel was normalized to the signal from the reference channel, resulting in increased stability and refractive index resolution of 5×10^{-5} RIU [69]. An integrated optical SPR sensor with wavelength modulation (Fig. 22) was demonstrated to provide a refractive index resolution as low as 1×10^{-6} RIU [70].

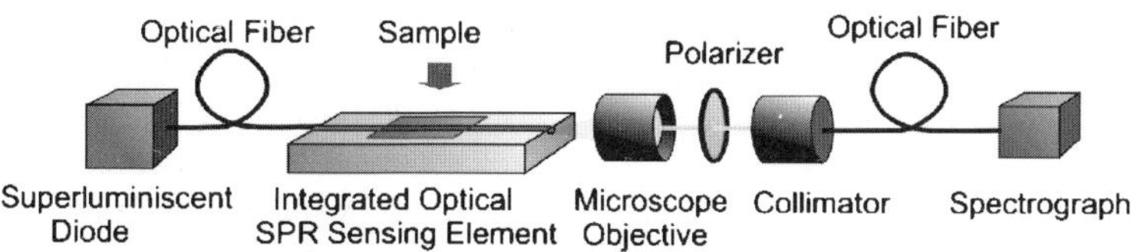

Fig. 22 Integrated optical SPR sensor with wavelength modulation [70]

The operating range of an integrated SPR sensor is determined by the refractive indices of the materials involved and the operating wavelength, which sets the operation range above 1.4 for conventional waveguides and wavelengths in the visible and near infrared. In order to shift the operating range of integrated optical sensors so that it includes aqueous environments, various approaches have been explored. These include an integrated optical waveguide fabricated in low refractive index glass [71], a buffer layer [67], a high refractive index overlayer [72], and more complex multilayer structures [73, 74]. However, all the approaches that introduced additional layers were found to yield less sensitive SPR sensing devices because of a relatively lower concentration of electromagnetic field in the sample.

4
Summary and Outlook

Over the last two decades, SPR instrumentation has made great strides in terms of optical systems, data processing, and (micro)fluidics. Bench-top high performance systems as well as small, compact SPR sensing devices have been realized. SPR sensors based on grating couplers have shown promise for mass production of low-cost SPR sensing devices. Highly miniaturized SPR sensors based on optical waveguides have also been demonstrated, although their fabrication is still rather complex and costly.

As SPR biosensor technology advances toward parallelized high-throughput screening systems, there is a growing need for optical systems compatible with massively parallel operations. SPR imaging presents a promising approach especially for applications where resolution is not a key issue. In the future, this approach is likely to be complemented by platforms based on spectroscopy of surface plasmon sensor arrays. Other directions for future research in the field of SPR sensors include improvement of the sensitivity of SPR sensor technology and development of robust referencing approaches, maintaining performance of SPR sensors even in non-laboratory environments.

References

1. Gordon JG, Ernst S (1980) Surf Sci 101:499
2. Nylander C, Liedberg B, Lind T (1982) Sensor Actuator 3:79
3. O'Brien MJ, Perez-Luna VH, Brueck SRJ, Lopez GP (2001) Biosens Bioelectron 16:97
4. Kawazumi H, Gobi KV, Ogino K, Maeda H, Miura N (2005) Sensor Actuator B Chem 108:791
5. Liedberg B, Lundstrom I, Stenberg E (1993) Sensor Actuator B Chem 11:63
6. Nenninger GG, Piliarik M, Homola J (2002) Measur Sci Technol 13:2038
7. Johansen K, Stalberg R, Lundstrom I, Liedberg B (2000) Measur Sci Technol 11:1630

8. Kukanskis K, Elkind J, Melendez J, Murphy T, Miller G, Garner H (1999) Anal Biochem 274:7
9. Goddard NJ, Pollardknight D, Maule CH (1994) Analyst 119:583
10. Thirstrup C, Zong W (2005) Sensor Actuator B Chem 106:796
11. Sjölander S, Urbanitzky C (1991) Anal Chem 63:2338
12. Stenberg E, Persson B, Roos H, Urbaniczky C (1991) J Colloid Interface Sci 143:513
13. Chinowsky TM, Jung LS, Yee SS (1999) Sensor Actuator B Chem 54:89
14. Tobiska P, Homola J (2005) Sensor Actuator B Chem 107:162
15. Piliarik M, Vaisocherova H, Homola J (2005) Biosens Bioelectron 20:2104
16. Berger CEH, Beumer TAM, Kooyman RPH, Greve J (1998) Anal Chem 70:703
17. Fu E, Chinowsky T, Foley J, Weinstein J, Yager P (2004) Rev Sci Instr 75:2300
18. Matsubara K, Kawata S, Minami S (1988) Appl Opt 27:1160
19. Lofas S, Malmqvist M, Ronnberg I, Stenberg E, Liedberg B, Lundstrom I (1991) Sensor Actuator B Chem 5:79
20. Karlsson R, Stahlberg R (1995) Anal Biochem 228:274
21. Nice EC, Catimel B (1999) Bioessays 21:339
22. http://www.biacore.com
23. Zhang HQ, Boussaad S, Tao NJ (2003) Rev Sci Instr 74:150
24. Song FY, Zhou FM, Wang J, Tao NJ, Lin JQ, Vellanoweth RL, Morquecho Y, Wheeler-Laidman J (2002) Nucl Acids Res 30:e72
25. Thirstrup C, Zong W, Borre M, Neff H, Pedersen HC, Holzhueter G (2004) Sensor Actuator B Chem 100:298
26. Pedersen HC, Thirstrup C (2004) Appl Opt 43:1209
27. Melendez J, Carr R, Bartholomew DU, Kukanskis K, Elkind J, Yee S, Furlong C, Woodbury R (1996) Sensor Actuator B Chem 35:212
28. Chinowsky TM, Quinn JG, Bartholomew DU, Kaiser R, Elkind JL (2003) Sensor Actuator B Chem 91:266
29. Kano H, Mizuguchi S, Kawata S (1998) J Opt Soc Am B Opt Phys 15:1381
30. Kano H, Knoll W (1998) Opt Comm 153:235
31. Stabler G, Somekh MG, See CW (2004) J Microsc 214:328
32. Kano H, Knoll W (2000) Opt Comm 182:11
33. Homola J, Pfeifer P, Brynda E (1997) Proc SPIE 3105:318
34. Brynda E, Homola J, Houska M, Pfeifer P, Skvor J (1999) Sensor Actuator B Chem 54:132
35. Homola J, Dostalek J, Chen SF, Rasooly A, Jiang SY, Yee SS (2002) Int J Food Microbiol 75:61
36. Nenninger GG, Clendenning JB, Furlong CE, Yee SS (1998) Sensor Actuator B Chem 51:38
37. Homola J, Lu HB, Yee SS (1999) Electron Lett 35:1105
38. Homola J, Lu HBB, Nenninger GG, Dostalek J, Yee SS (2001) Sensor Actuator B Chem 76:403
39. Dostalek J, Vaisocherova H, Homola J (2005) Sensor Actuator B Chem 108:758
40. Cahill CP, Johnston KS, Yee SS (1997) Sensor Actuator B Chem 45:161
41. Akimoto T, Ikebukuro K, Karube I (2003) Biosens Bioelectron 18:1447
42. Nenninger GG, Tobiska P, Homola J, Yee SS (2001) Sensor Actuator B Chem 74:145
43. Slavík R, Homola J (2006) Opt Comm 259:507
44. Rothenhausler B, Knoll W (1988) Nature 332:615
45. Hickel W, Kamp D, Knoll W (1989) Nature 339:186
46. Bassil N, Maillart E, Canva M, Levy Y, Millot MC, Pissard S, Narwa W, Goossens M (2003) Sensor Actuator B Chem 94:313

47. Fu E, Foley J, Yager P (2003) Rev Sci Instr 74:3182
48. Zybin A, Grunwald C, Mirsky VM, Kuhlmann J, Wolfbeis OS, Niemas K (2005) Anal Chem 77:2393
49. Nikitin PI, Grigorenko AN, Beloglazov AA, Valeiko MV, Savchuk AI, Savchuk OA, Steiner G, Kuhne C, Huebner A, Salzer R (2000) Sensor Actuator A Phys 85:189
50. Nikitin PI, Beloglazov AA, Kochergin VE, Valeiko MV, Ksenevich TI (1999) Sensor Actuator B Chem 54:43
51. Notcovich AG, Zhuk V, Lipson SG (2000) Appl Phys Lett 76:1665
52. Dostalek J, Homola J, Miler M (2005) Sensor Actuator B Chem 107:154
53. Jory MJ, Vukusic PS, Sambles JR (1994) Sensor Actuator B Chem 17:203
54. Jory MJ, Bradberry GW, Cann PS, Sambles JR (1995) Meas Sci Technol 6:1193
55. Adam P, Dostalek J, Homola J (2006) Sensor Actuator B Chem 113:774
56. Adam P (2005) Surface plasmons on diffraction gratings and their sensor applications. Charles University, Prague
57. Brockman JM, Fernandez SM (2001) Am Lab 33:37
58. Jorgenson RC, Yee SS (1993) Sensor Actuator B Chem 12:213
59. Jorgenson RC, Yee SS (1994) Sensor Actuator A Phys 43:44
60. RonotTrioli C, Trouillet A, Veillas C, Gagnaire H (1996) Sensor Actuator A Phys 54:589
61. Bender WJH, Dessy RE, Miller MS, Claus RO (1994) Anal Chem 66:963
62. Homola J (1995) Sensor Actuator B Chem 29:401
63. Slavik R, Homola J, Ctyroky J (1998) Sensor Actuator B Chem 51:311
64. Slavik R, Homola J, Ctyroky J, Brynda E (2001) Sensor Actuator B Chem 74:106
65. Piliarik M, Homola J, Manikova Z, Ctyroky J (2003) Sensor Actuator B Chem 90:236
66. Van Gent J, Lambeck PV, Kreuwel HJM, Gerritsma GJ, Sudholter EJR, Reinhoudt DN, Popma TJA (1990) Appl Opt 29:2843
67. Lavers CR, Wilkinson JS (1994) Sensor Actuator B Chem 22:75
68. Mouvet C, Harris R, Maciag C, Luff B, Wilkinson J, Piehler J, Brecht A, Gauglitz G, Abuknesha R, Ismail G (1997) Anal Chim Act 338:109
69. Harris RD, Luff BJ, Wilkinson JS, Piehler J, Brecht A, Gauglitz G, Abuknesha RA (1999) Biosens Bioelectron 14:377
70. Dostalek J, Ctyroky J, Homola J, Brynda E, Skalsky M, Nekvindova P, Spirkova J, Skvor J, Schrofel J (2001) Sensor Actuator B Chem 76:8
71. Harris RD, Wilkinson JS (1995) Sensor Actuator B Chem 29:261
72. Ctyroky J, Homola J, Skalsky M (1997) Electron Lett 33:1246
73. Weiss MN, Srivastava R, Groger H (1996) Electron Lett 32:842
74. Weiss MN, Srivastava R, Groger H, Lo P, Luo SF (1995) Sensor Actuator A Phys 51:211
75. Homola J, Vaisocherová H, Dostálek J, Piliarik M (2005) Methods 37:26

Springer Ser Chem Sens Biosens (2006) 4: 117–151
DOI 10.1007/5346_017

Published online: 8 July 2006

The Art of Immobilization for SPR Sensors

Stefan Löfås (✉) · Alan McWhirter

Biacore AB, Rapsgatan 7, 75450 Uppsala, Sweden
Stefan.lofas@biacore.com

Keywords Biacore · Biosensors · Immobilization and surface chemistry · Protein chips · Protein interaction analysis · Surface plasmon resonance

1
Introduction

Since the first seminal work on the use of SPR-based detection technology in bioanalytical applications [1], the field has seen a tremendous development.

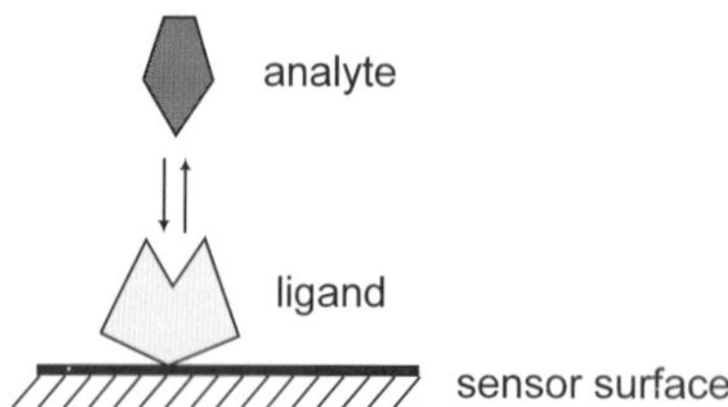

Fig. 1 One interaction partner is immobilized to the sensor surface. The analyte is free in solution and binds to the immobilized molecule (denoted "ligand" in this and subsequent figures)

This is manifested in numerous ways, including the growing use of SPR in research as well as in the many significant improvements in commercial instruments, which have opened up their use for a wide range of applications and user groups. Technical advances have been made in many areas, including the detection unit, fluid and sample handling, data treatment and not least, in the immobilization procedures for functionalization of the sensor surface. This chapter will deal with the progress made in surface modification techniques and approaches for immobilizing interacting partners on these surfaces (Fig. 1).

Although immobilization on solid surfaces or matrices has been described and practiced for several decades, SPR-based biosensors pose some unique requirements. A successful direct, label-free measurement of specific binding events will be facilitated by the best possible activity of the immobilized interactant. It is a general rule that all types of non-specific binding to the surface must be kept as low as possible in order to prevent irrelevant signals interfering with the interpretation of the specific interaction. Since SPR detection can be applied to a great variety of analytical applications, a correspondingly large range of methods for immobilization have been developed. Given the wide variation in molecular properties, no generally applicable immobilization method has emerged. Rather, even among proteins, different approaches may be needed in order to reach the required activity. Approaches for both covalent immobilization and for affinity-based capture methods will be reviewed.

The SPR phenomenon is ideally suited for miniaturization and for array format applications. Methods for the immobilization of the range of molecules that can be expected for array formats have also been developed and implications and issues related to this will be described. Finally, future trends and opportunities related to immobilization for SPR detection will be discussed.

2 Surface Modifications

Early descriptions of SPR technology for bioanalytical applications were based on simple physical adsorption of proteins to an active metal surface [1]. However, it was soon realized that a more sophisticated approach was needed in order to meet the challenges demanded by the range of potential applications involved. Commonly used metal substrates such as gold and silver show a high tendency for spontaneous adsorption of proteins and other molecules. This passive binding to the metal substrate results in a loss of the bioactivity. Similarly, studies on antibody binding activities in ELISA-type assays after their adsorption to plastic surfaces have shown levels as low as 2–10% of the adsorbed amount [2].

These effects can be explained by a reorganization of the immobilized molecule to attain the most favorable thermodynamic state. For example, adsorption to hydrophobic surfaces is driven by rearrangements that optimize contact of hydrophobic segments with the substrate. Passive binding to a surface substrate also opens possibilities for uncontrolled exchange of the immobilized molecule during an analysis cycle. If the modified surface is used for repeated analysis cycles, the probability of exchange will be further enhanced and lead to unreliable assays.

2.1 Coating of Surfaces with Self-Assembled Monolayers

Extensive efforts have been made to develop approaches for coating metal surfaces before immobilization. This serves to minimize non-specific adsorption, as well as to introduce reactive groups for specific immobilization. The most successful methods are based on the concept of molecular self-assembly of thiol- or disulfide molecules on the metal surface. The spontaneous formation of organic disulfide monolayers on gold was initially shown by Allara and Nuzzo in 1983 [3] in the context of models for interface studies. Monolayer formation is driven by a strong coordination of sulfur with the metal, accompanied by van der Waals interactive forces between the alkyl chains. With a sufficient chain length, the resulting monolayer forms a densely packed and very stable structure that is oriented more or less along the normal to the metal surface (Fig. 2).

These nanometer-thick layers are easily fabricated from commercially available substances, or can at least be synthesized with relative ease [4]. The first applications of self-assembled monolayers (SAM) for biosensor use were described in the late 1980s and originally developed for Biacore instruments [5, 6]. Hydroxyl-terminated long chain thiol alkanes were designed for the formation of the SAM on gold. Such layers can be activated for direct linkage of various molecules or further derivatized with different chemistries

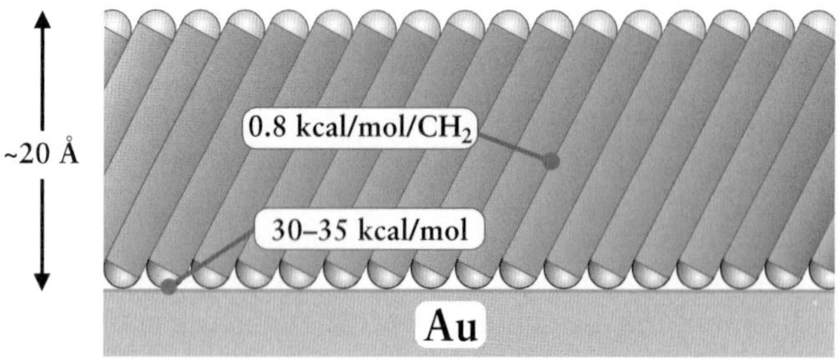

Fig. 2 Schematic illustration of a self-assembled monolayer structure on a gold substrate

for more advanced surface modifications, as will now be described in more detail.

The possibility for different functional end groups in the alkyl thiols creates a high degree of flexibility in terms of the types of surface properties that can be obtained. Extensive studies involving various types of coatings have been reviewed elsewhere [4, 7, 8]. For example, in early applications [5, 6], a terminal hydroxyl function was introduced to give the surface a highly hydrophilic character, while acting as a means for immobilization of various molecules, either directly or via suitable linkers. Direct covalent immobilization of proteins to various ω-terminated groups has also been described, although there are limitations with such approaches.

2.2 Development of the Dextran Hydrogel

Even if flat surface substrates are made hydrophilic, their rigid character may induce denaturation or impaired activity of proteins [9]. Furthermore, SPR senses mass-dependent refractive index changes a few hundred nanometers from the metal surface. Taking advantage of this, surface modification procedures were developed for sensor chips produced for the company, Biacore AB in which a thin hydrogel-like polymer layer based on dextran was introduced. The dextran polymer is composed of mainly unbranched glucose units, providing high flexibility and water solubility. Immobilization is facilitated via epoxy modification of the terminal hydroxyl SAM and subsequent nucleophilic reaction of the dextran under alkaline conditions [5, 6, 10] (Fig. 3).

The surface can be further activated with suitable linkers for subsequent immobilization, and here the introduction of carboxymethyl groups has proven a versatile alternative. By choosing different sizes of dextran, ranging from 10 kDa to over one million Da, surfaces tailored for specific applications can be created. This type of surface modification serves multiple purposes. The hydrogel-like layer provides a highly hydrophilic environment.

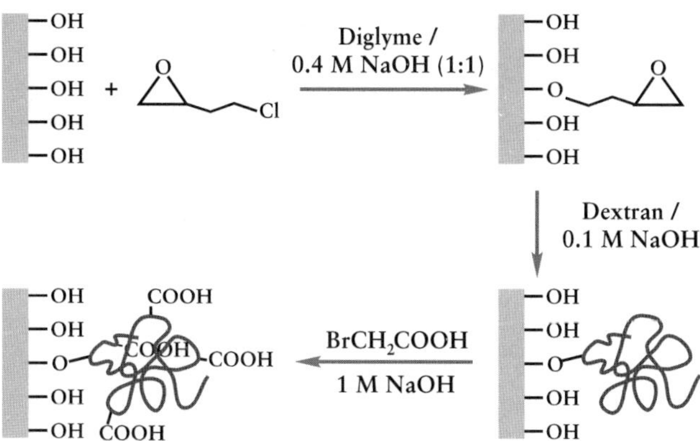

Fig. 3 Synthesis sequence for the construction of a carboxymethyl dextran-coated sensor surface

The glucose-based polymer is highly suited for well-defined covalent immobilization of proteins that rely on a wide variety of chemistries. Furthermore, the extended matrix structure has been shown to increase the binding capacity several-fold compared to flat surfaces. Finally, this thin layer extension is well matched with respect to the penetration depth of the evanescent wave [11, 12].

The linkage of dextran polymer chains to the sensor surface provides an open, non-cross-linked structure on which immobilized molecules can attain a solution-like state with a certain level of freedom to move around within the hydratized layer. This view is supported by the excellent agreement that has been obtained in comparisons of affinity data from Biacore's SPR-based platforms and solution-based methods [13]. The most commonly used carboxymethylated derivate of dextran surfaces also have the benefit of improved solubility properties. The degree of carboxymethyl modification can be modulated for different applications and sensor surface capacity requirements. These types of negatively charged layers may exhibit electrostatic background binding of basic compounds, which needs to be considered in the design of the immobilization procedure and the assay. However, working under physiological buffer conditions normally suppresses such effects by electrostatic shielding. Alternatively, lowering the degree of carboxymethylation can also be used to reduce this effect.

Other hydrophilic polymers have been conceived as alternatives to dextran [6]. For example, polyvinyl alcohol and polyacryl acid derivatives are feasible and graft combinations thereof have been shown to be applicable to SPR detection [14]. Poly-L-lysine has become popular for DNA microarray coatings, due to its highly positive charge. It has also been attached to SAM-derivatized gold surfaces for subsequent modification with thiol reactive groups [15].

2.3 Further Chemical Modifications to Optimize the Sensor Surface

Although flat or two-dimensional (2D) surfaces are used for various applications within the DNA and protein microarray area, practical uses for SPR detection were originally limited. This can be attributed to the sensitivity limitations of the technology, despite the relative ease in handling and the wide variety of developed chemistries based on flat surface structures. Thus, an immobilized monolayer may not give sufficient binding responses under certain conditions, especially if immobilization leads to compromised activity of the immobilized partner. As described previously, non-specific binding also needs careful consideration.

These potential limitations, however, have been largely eliminated. Early attempts describe modifications of the metal surface with thin insoluble layers, such as silica with subsequent functionalization via silane compounds [16]. The SAM approach created a tool for convenient introduction of various surface functionalities that can be used for immobilization [6]. Examples include SAMs that are ω-terminated with hydroxyl or carboxyl groups, which can be activated for covalent coupling via nucleophilic reactions. In addition, this modification with epoxy groups leads to activated surfaces that can be directly used for nucleophilic linkage.

A similar approach has been developed for biotin-based surfaces that can be further modified using streptavidin. Such structures provide a general capture tool by binding a wide variety of biotinylated compounds. Knoll et al. developed SAM-based surface modifications where ω-terminated biotinylated alkane thiols were utilized in different forms [17]. By mixing biotinylated molecules with hydroxyl-terminated thiols as diluting agents in different ratios, a 1 : 9 molar ratio was found to be optimal for binding monolayers of streptavidin. This is in contrast to a SAM composed of a single biotinylated thiol, where the biotin groups are sterically hindered from binding to streptavidin. By utilizing different alkyl chain lengths in the biotinylated thiols and the diluting molecules, the biotins can be exposed to more efficient binding of streptavidin. This strategy can also be used for other functional groups, such as combinations of carboxy- and hydroxyl-terminated thiols. The diluted biotin surfaces can also be generated by reaction of biotin derivatives with suitably functionalized SAMs, modifying a fraction of carboxy-terminated thiols with amine derivatives of biotin [18]. Both approaches have their limitations; the use of biotinylated or other modified thiols may be limited by the accessibility and cost of such molecules, while the surface modification strategy can be difficult to apply consistently.

When implemented correctly, however, both approaches yield streptavidin surfaces with good binding capacity and generally sufficient biocompatibility, even if there are reports that streptavidin has a tendency for unwanted binding of a range of compounds [19].

Several alternatives to biotin for mixed functionalized SAMs have been described. One of the more interesting approaches involves the use of short oligo ethylene glycol (OEG) units to increase the biocompability and to extend the functional alkyl thiol [8, 20]. Whitesides et al. have described mixed SAMs for use in SPR detection composed of *N*-hydroxysuccinimide activated carboxy-terminated OEG thiol alkanes and shorter hydroxyl-terminated analogs [21]. The OEG spacers constitute what can be considered as a very thin hydrogel like layer, creating surfaces with properties that resemble both 2D and 3D layers. The suppression of non-specific binding is similar or better than dextran-based surfaces, but capacities are limited to monolayer levels similar to other 2D surfaces. Extensive studies have investigated protein resistance effects by OEG-terminated SAMs, and these indicate that binding of interfacial water by the OEG moieties is important [22]. The susceptibility for oxidation and degradation of OEG-based structures may also present a practical problem, limiting storage stability and performance quality (see footnote 17 in [22]). Alternative spacing units that overcome these limitations have been evaluated and reviewed elsewhere [8, 23].

3 Immobilization Techniques

The development of SPR technology has encouraged the development of numerous strategies for the immobilization of different types of recognition elements. These have focused on proteins but also include methods for peptides, DNA, RNA, carbohydrate structures and organic molecules of various kinds including lipids and more complex natural products.

Immobilization methods have been successively developed from earlier adsorption processes, using highly controlled general covalent chemistries and specific alternatives of various kinds. Considerable experience has also

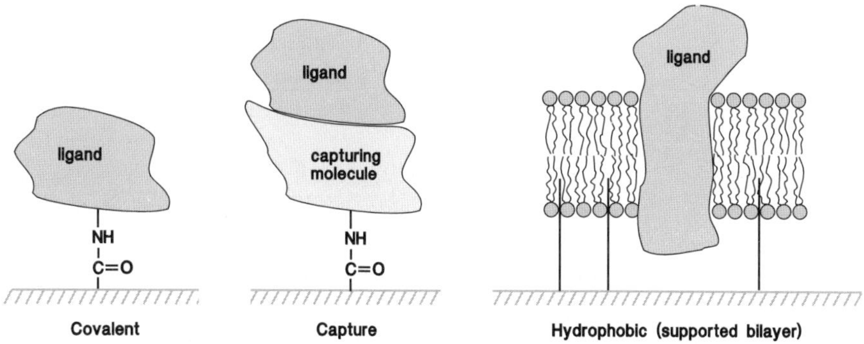

Fig. 4 Different approaches for immobilizing binding partners to the sensor surface

been gained from immunochemistry and affinity chromatography, where issues related to the maintenance of activity are also relevant [24, 25]. However, the miniaturization of the sensor areas and the true heterogeneous, interfacial conditions needed for immobilization are factors that have required novel solutions for satisfactory results. The availability of these methods has also been an important factor for the acceptance of SPR-based instrumentation as an established and widespread analytical tool.

The following section describes the most important immobilization techniques, including different covalent coupling alternatives, non-covalent capture techniques and more specialized methods for lipids and membrane proteins (Fig. 4).

3.1 Covalent Immobilization

Limitations in simple adsorption processes have led to the development of advanced surface coatings designed for controlled immobilization. Different functional groups have been introduced on the surface, enabling the formation of a stable linkage to another appropriate functional group. This may include an activation step of one or both of the functional groups, which results in transformation into a more reactive form. For proteins in particular, the chemistries utilized also need to be performed under relatively mild conditions and in aqueous solutions, placing certain limitations on the available repertoire.

The possibility of having a transformable functional group on the sensor surface is an attractive concept as a general starting surface for use with a range of coupling chemistries. The carboxymethylated dextran coating described in the previous section was designed to include the carboxylic acid residue as a functional group that can be used either for direct coupling, or

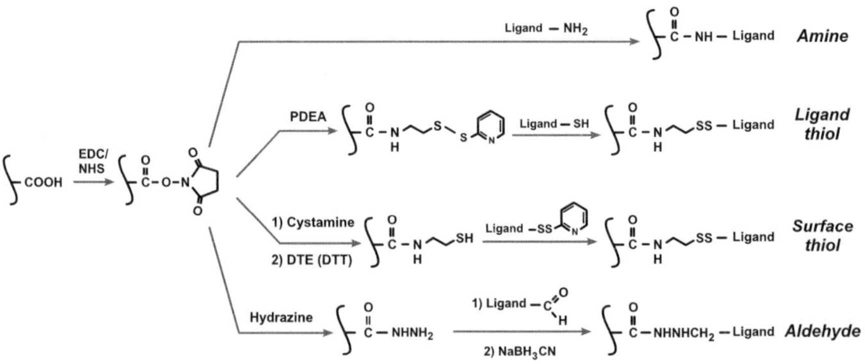

Fig. 5 Reaction sequences for different binding partner immobilizations based on couplings to the carboxylic acid group

switched to other functionalities. Figure 5 shows how the carboxylic groups can either be directly reacted with amine groups or converted for use in coupling chemistries based on thiol reactions, aldehyde and carboxylic acid condensations, and biotin capture techniques.

More detailed descriptions of the various covalent couplings will be given in the following section. Notably, a literature review indicates that the carboxymethyl dextran surface used in Biacore instruments in combination with the amine coupling method is by far the most widely used immobilization strategy [26].

3.1.1
Coupling of Nucleophiles to Carboxylic Groups

The most versatile and widely used approach involves coupling with reactive nucleophile functionalities to carboxylic groups on the sensor surface. The most common nucleophile utilized is the amine group in lysine residues, but hydroxyl groups can also be used. The carboxylic groups are readily introduced on dextran or other hydroxyl-containing surface layers via reactive haloacetic acids. In the case of SAM layers, alkane thiols that are ω-terminated with carboxylic groups can be utilized. To achieve the formation of a covalent amide or ester bond between the carboxylic and amine or hydroxyl groups, respectively, activation with carbodiimide reagents is most commonly used. Water-insoluble carbodiimides such as DCC (dicyclohexyl carbodiimide) are normally used in organic chemistry applications, but for reactions in aqueous solutions, alternatives like EDC (1-ethyl-3-(3-dimethylaminopropyl) carbodiimide) are preferred [27]. The purpose of the carbodiimide reagent is to create a reactive O-acyl isourea intermediate with the carboxylic group, which is then reacted with a suitable nucleophile. The coupling is normally performed in two steps, with activation followed by reaction, in order to avoid reaction between the carbodiimide and the immobilized molecule. However, in aqueous solutions the reactivity of the intermediate is so high that water hydrolysis rapidly transforms it back to carboxylic acid, if it is not trapped by another competing nucleophile. This side effect is conveniently overcome using a mixture of the carbodiimide and a reactive hydroxyl compound, forming an active ester derivative that is stable for several minutes to hours (Fig. 6).

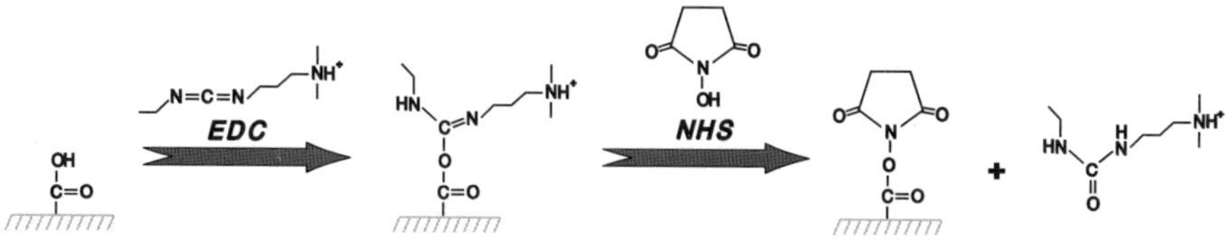

EDC NHS

Fig. 6 Activation sequence of the carboxylic acid group with EDC/NHS

N-Hydroxy succinimide (NHS) has been found to be a very suitable reagent for these purposes and is normally mixed at high concentrations with EDC in water. Both the EDC and NHS act as buffering agents and a pH around 5–6 is obtained, providing conditions for an optimal reaction rate for NHS ester formation. Although other ester-forming compounds like nitrophenol and its derivatives are also possibilities, NHS is normally preferred due to its solubility in water, relatively low toxicity, and optimal reactivity for two-stage couplings. Extensive optimization studies have been performed on the activation and coupling conditions for protein immobilization to the carboxymethylated sensor surfaces developed by Biacore [27].

Coupling to the active esters can be carried out under various conditions depending on the molecular type. Displacements in aqueous solutions are normally done under slightly alkaline conditions, e.g., in carbonate or borate buffers around pH 8.5, where a normal alkylamine nucleophile is close to its pKa and can compete with water hydrolysis. This is also the preferred method for organic molecules and small peptides. These conditions have also been widely practiced when immobilizing proteins for affinity chromatography [24]. An alternative approach was developed for in situ immobilization of proteins to sensor surfaces [5, 6] where high-density modifications are desirable and has now become the standard method of choice. This concept relies on electrostatic attraction of the proteins to an NHS-activated carboxylated surface, on which a fraction of the carboxylic groups remain unreacted. Under low ionic strength buffer conditions, where the surface is negatively charged and the protein has a positive charge, a high local surface concentration of the protein is obtained. This greatly favors a reaction of the nucleophiles on the proteins over water hydrolysis of the esters. Suitable buffer conditions to achieve this are normally obtained by working in 10 mM acetate buffers at pH 4–6, where a large fraction of all proteins are positively charged [27]. Much lower protein concentrations than those normally used in coupling to solid phases can consequently be employed. The reaction times are also considerably shorter, in the range 1–10 min.

The electrostatic attraction approach can be applied to all types of surfaces that have a combination of reactive groups and residual charges. The most successful implementations, however, are found for 3D surfaces such as carboxymethylated dextran. Here, the attraction leads to multilayers of bound protein. Quantifications using radioactively labeled proteins that were also used to calibrate the SPR responses showed surface concentrations of up to 50 ng mm^{-2}, which represents several high-density packed monolayers for typical proteins [28]. Furthermore, covalent coupling occurs under very mild conditions, where only a small fraction of the nucleophilic groups on the protein are reactive (e.g., the amino groups on the lysine residues are unprotonated). This leads to very few immobilization points, little or no cross-linking and a high likelihood of preserving activity. Protein A immobilized under these conditions showed a binding capacity of more than three IgG

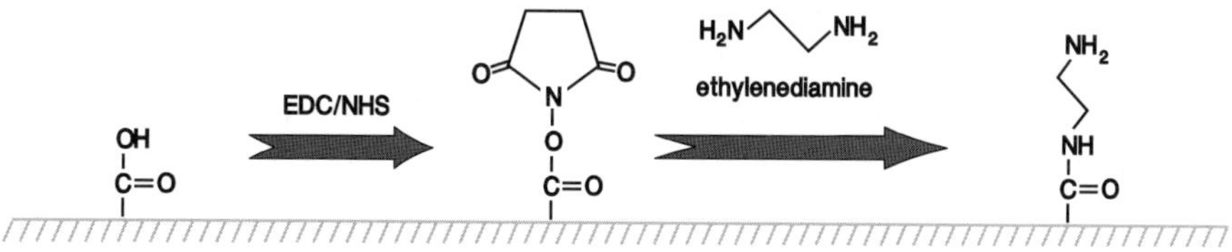

Fig. 7 Modification of the sensor surface with amine groups via EDC/NHS activation and ethylene diamine reaction

molecules per protein A molecule [27]. Likewise, immobilized IgG antibody molecules showed antigen binding capacities approaching 1.5 antigens per antibody (75% activity based on two antigen binding sites per antibody) [29]. These results stand in sharp contrast to results reported for immobilizations of monoclonal antibodies to chromatography supports, where low activities in the range 1–30% were obtained [30].

In some instances, the reverse approach may be preferred, with functional amine groups on the sensor surface and activated carboxylic groups. This strategy can be used when the molecule lacks appropriate reactive amines or other nucleophiles, or when the nucleophile is suspected to be close to the analyte binding site. The approach is particularly valuable when working with small organic molecules, as will be further described in Sect. 4.4. Amine groups can be introduced to the sensor surface in several ways. A convenient route involves the conversion of carboxylated surfaces via EDC/NHS activation and subsequent ethylene diamine reaction (Fig. 7).

Surfaces functionalized with primary amine groups should normally be further reacted directly after they are produced, as the amine groups rapidly lose reactivity when kept in normal aqueous buffer conditions [31]. This is believed to occur by carbamate formation via reaction with carbon dioxide, and also via oxidation phenomena.

3.1.2 Couplings to Thiol Groups

Although amine coupling to activated carboxylic groups is the most commonly used form of covalent immobilization, there are alternative approaches that may be preferable under certain circumstances. Amine coupling may occur at or near the active site, or the molecule may lack amine groups (which may not be possible to introduce due to chemical restrictions). A useful alternative is thiol coupling, which relies on reactive functionalities that are thiol-selective [32]. The thiol reactive groups most commonly used are active disulfides such as pyridyl disulfides or their derivatives, although maleimide and acyl halide derivates are common alternatives. The thiol groups can either be introduced on the sensor surface and reacted with molecules with thiol reactive groups (Fig. 8), or performed in reverse,

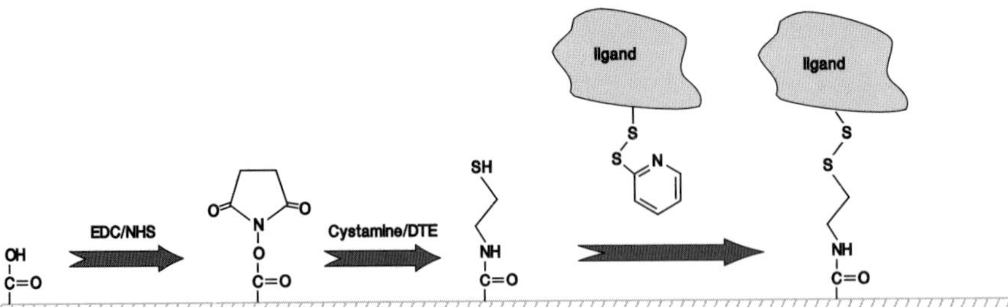

Fig. 8 Coupling of binding partners to thiol-modified sensor surface

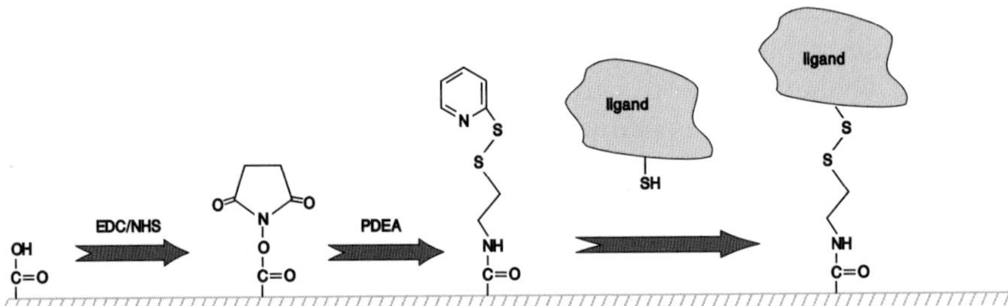

Fig. 9 Coupling of thiolated binding partner to pyridyl disulfide-modified sensor surface

with the active group on the surface and the thiol on the molecule to be immobilized (Fig. 9).

Various reagents for modifications based either on reactive disulfides, maleimides, or acyl halides are commercially available. One advantage of modifying a protein is the possibility of minimizing the number of coupling sites in order to avoid blocking the active site or to keep cross-linking low. Thus, even if the reagents are frequently directed toward amine groups, reaction conditions may be steered to preserve activity.

The disulfides can be coupled under very mild conditions, and in the case of pyridyl disulfides, even in acidic buffers [33]. The selectivity is also very high, with little or no interference from other nucleophiles. The disulfide bond can undergo exchange reactions with free thiol compounds that may limit stability under certain conditions. For example, buffers with added thiols such as mercaptoethanol may induce disulfide bond cleavage and dissociation of the immobilized binding partner. This effect has also been exploited for the reuse of modified surfaces. After cleavage with a reactive thiol under mildly alkaline conditions, the residual thiol groups on the sensor surface can be used for immobilizing disulfide-containing molecules [34].

The maleimide and acyl halide reaction proceeds via Michael addition and forms a thioether linkage to the thiol that is normally more stable than the disulfide bond (Fig. 10). The thioether bond is normally formed at pH 7.5

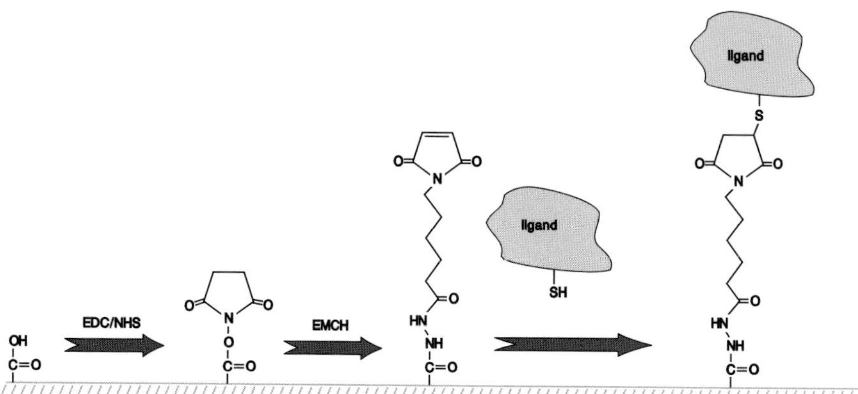

Fig. 10 Coupling thiolated binding partner to maleimide-modified sensor surface

to 8.5 but with somewhat lower selectivity than the disulfide reaction. Competition from other nucleophilic groups can occur under certain conditions and this needs to be considered in the choice of immobilization.

Coupling methods exploiting thiol groups can also be performed under electrostatic concentration conditions similar to those described for amine coupling [32]. As this step is normally carried out under acidic conditions, the method is best performed using reactive pyridyl disulfides.

3.1.3 Coupling to Aldehyde Groups

Schiff base condensation of aldehyde groups to amines and hydrazides has been utilized for glycoprotein immobilizations in chromatography applications [2, 35]. This method exploits the generation of aldehyde functionality by oxidation of carbohydrate residues in proteins. Standard protocols are available for mild oxidation of the sugars using sodium periodate solution. Sialic acid residues in particular, readily form aldehydes by cleavage of the exocyclic vicinal diol. The carbohydrates are not normally located near the active site and consequently, the resulting site-specific coupling may yield immobilized molecules with high binding activity. IgG antibodies in particular are well suited for aldehyde mediated immobilization.

Although Schiff base formation can be performed with amine groups, the low stability of the bond in aqueous conditions makes hydrazide a better alternative. Hydrazides can be introduced on the sensor surface via reaction of hydrazine or carbohydrazine to carboxylic groups after activation with EDC/NHS (Fig. 11) [32]. The hydrazide–aldehyde bond forms rapidly and is relatively stable in neutral to alkaline conditions, but disintegrates slowly in acidic buffers. If necessary, the bond can be further stabilized by reduction with sodium cyanoborohydride at pH 4.

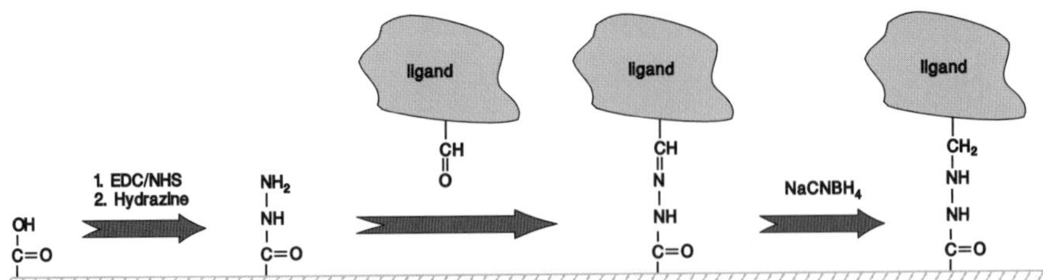

Fig. 11 Coupling of aldehyde containing binding partners to hydrazide-modified sensor surface, followed by cyanoborohydride reduction

As with covalent coupling methods, aldehyde coupling is well suited to electrostatic concentration conditions, provided the sensor surface holds residual negative charges.

In contrast to observations made in chromatography, the "site-specific" immobilization of antibodies via aldehyde groups to carboxymethyl dextran surfaces does not significantly improve activities [32]. This is probably related to the mild conditions that can be utilized in amine coupling (Sect. 3.1.1), which minimize multisite immobilization and thus preserve the binding activity, in combination with immobilization to the extended dextran polymer layer. Aldehyde coupling may be a good alternative for smaller proteins, as there may be a higher probability of masking the binding site through random coupling.

3.2 Capture-Based Coupling of Native and Tagged Molecules

As indicated in Sect. 3.1, covalent coupling techniques are limited under certain situations, for example, the molecule may be unstable under the required coupling conditions or the activity of the binding site may be impaired. Problems may also arise if the molecule is present in small amounts in cell lysates or other complex sample matrices. Immobilization based on non-covalent capture may be a good alternative in such cases. Capture is based on high affinity binding via a specific integral region or recombinant tag to a capturing agent on the sensor surface. An additional benefit of this approach is the possibility for removal of the immobilized binding partner after the analysis by an analyte-independent regeneration step, followed by renewed capture in the next assay cycle. The disadvantage with this alternative compared to covalent coupling is the significantly increased sample consumption. In addition, for applications with demands on quantitative data, the affinity must be sufficiently high that dissociation is insignificant. This can be achieved by the introduction of multiple tags to increase the strength by avidity binding to the capture agent on the sensor surface. A disadvantage here is that this may lead to decreased activity due to interference with the binding site and/or cross-linking. For SPR detection, therefore, it is usually recommended to keep the degree of modification as low as possible.

The most commonly used capturing agents are specific antibodies directed towards tagged recombinant proteins. Frequently used tags include GST, Myc, FLAG, and poly-His residues. Integral residues in proteins may also be addressed, particularly for the capture of monoclonal antibodies (mAbs) from growth media. Here, antibodies specific for the Fc region of the mAb may be used. Other alternatives involve the use of protein A or protein G for selective IgG antibody capture.

The capture molecule is most frequently a protein (particularly an antibody), but can also be composed of organic molecules. His-tagged proteins can be selectively captured via the metal-chelating complex based on nitrilotriacetic acid (NTA) and nickel ions. NTA derivatives are immobilized to the sensor surface, either via coupling to carboxymethyl dextran [36, 37] or by use of SAM approaches [38]. The bond can easily be broken with a solution of a chelating agent like EDTA and the sensor surface is then reused after activation with a Ni^{2+} solution. The intrinsic affinity in the Ni/NTA–His bond, however, is relatively low (μM) [39] and this may be insufficient for robust assay performance. Variations of this method have therefore been developed, combining affinity-based NTA capture and covalent coupling via amine groups to NHS-activated carboxyl groups on the sensor surface [40].

Alternative capture systems have recently been described that involve peptide–peptide interactions. A heterodimeric coiled-coil peptide domain can be utilized by conjugation of an E-coil strand to the protein and reversible immobilization to the K-coil strand coupled to carboxymethylated dextran surfaces [41].

A special case of the capture approach involves the use of the avidin–biotin affinity bond. This very popular conjugation method is widely used and is also highly suitable for SPR sensors. In addition to their use as a linkage between biotinylated SAM layers and biotin-modified molecules [17], streptavidin and other avidin variants can also be conveniently immobilized to carboxymethylated dextran via amine coupling [32]. Biotinylation reagents of various kinds are commercially available, together with protocols for optimal modification. The high affinity of the biotin–avidin bond (10^{-12}–10^{-15} M) makes it practically impossible to break without destroying the immobilized avidin molecule and should therefore rather be considered as a covalent bond in its behavior. Chemical variants of the biotin structure and recombinant versions of avidin have been developed in order to diminish the high binding strength, but these are correspondingly less robust under capture assay conditions.

Antibody-based capture agents have also been used for both the His-tags and biotinylated molecules. Commercial antibodies are available, but in practice no general-purpose reagents that are optimal for all tagged proteins have been found. Depending on the type of molecule and method of tag introduction, the binding strength is affected on a case-by-case basis, and different antibodies may need to be tested to obtain the best performance.

Promising new approaches include the specific and covalent surface immobilization of fusion-tagged proteins. One recent example utilizes a fusion tag composed of a mutant of the human DNA repair protein O^6-alkylguanine-DNA alkyltransferase (hAGT) [42]. A derivative of O^6-bensylguanine was immobilized to carboxymethylated dextran surfaces and selective coupling of the fusion protein was obtained. Specific immobilization directly from crude cell extracts that expressed the hAGT fusion protein was also demonstrated.

3.3 Coupling Mediated via Lipid Layers

The interest in SPR-based detection of proteins interacting with lipids or in a lipid environment has steadily increased, particularly as many membrane-associated proteins are drug target candidates. Lipid membranes themselves can also be targeted, e.g., in the development of antibiotically targeted drugs. Lipids, being amphiphilic and normally without functional groups for covalent immobilization, are difficult to immobilize but approaches have been developed to overcome this problem. These approaches are based on the adsorption of lipid vesicles or liposomes to certain types of surfaces. In contact with planar surfaces, liposomes tend to unfold and create a well structured and densely packed lipid monolayer, in which the hydrophobic part of the molecule is oriented perpendicularly towards the surface [43]. The lipid head group faces towards the aqueous solution and can interact with analytes. The structure is sufficiently stable for SPR detection, but can easily disintegrate in the presence of detergents. Both hydrophobic and hydrophilic sensor surfaces have been shown to work, supporting the formation of SAMs from long-chain alkane thiols [44, 45]. The hydrophobic surfaces are particularly sensitive, however, to disturbances from minor impurities in the solutions used. The impurities can adsorb and interfere with the liposomes and care must be taken to obtain reliable results.

Alternative surface modifications have therefore been developed for the formation of lipid bilayers. Phospholipids modified in the head group with hydrophilic thiolated spacers have been utilized for anchoring lipid bilayers to gold surfaces [46]. Gold surfaces with hydrophilic polymers have also been modified with hydrophobic groups to which liposome structures can tether [47]. A surface based on carboxymethylated dextran modified with long alkyl chains was developed and provides a convenient support for immobilization of lipids [48]. Depending on the type of lipid and liposome preparation conditions, either intact liposomes or planar lipid bilayers are formed in contact with these types of surfaces [49]. Methods have also been developed for rapid and controlled formation of planar bilayers in flow-based systems [50]. As shown in Fig. 12, a mixture of lipids and detergents is first

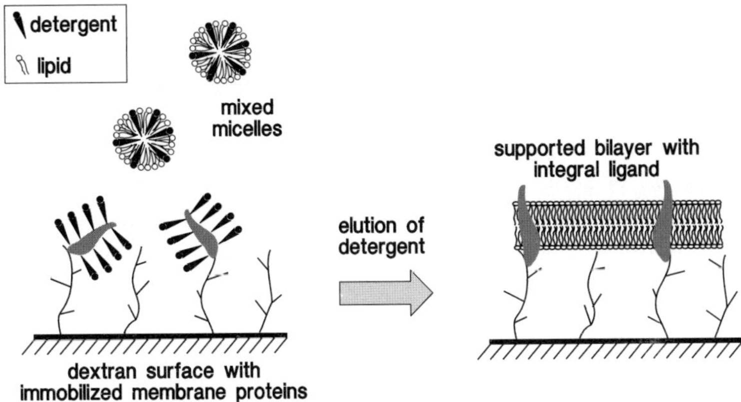

Fig. 12 On-surface reconstitution of immobilized membrane proteins with lipid bilayers by injection of mixed micelles and rapid detergent elution

injected, followed by a switch to pure buffer that depletes the detergent and leaves the lipid layer.

Alternative approaches for liposome assemblies have also been demonstrated, where histidine-tagged lipids have been introduced in vesicles, which were then anchored to chelator surfaces [39, 51]. Similarly, oligonucleotide-modified lipids can be incorporated in the vesicles and bind to complementary sequences immobilized on the sensor surface [52, 53]. The latter strategy can also be utilized for spatially resolved immobilizations.

Lipid bilayer surfaces are suitable for incorporation of membrane-associated protein receptors such as G protein-coupled receptors (GPCRs). This can be achieved by tethering vesicles with reconstituted receptors to the surface. Alternatively, solubilized proteins can be bound to the sensor surface followed by a rapid in-situ reconstitution by the lipid–detergent method [50] (Fig. 12). This method has the potential to produce receptor densities that are sufficient for use in SPR sensors. However, the widespread use of SPR in this field is still limited by the availability of membrane-associated proteins and their low stability, as well as by a lack of methods for handling them in a purified format.

3.4 Creating and Validating Functional Sensor Surfaces: General Comments and Practical Tips

Reliable SPR-based assays require solid foundations. In particular, it is important that the immobilization chemistry selected to couple a protein to the sensor surface does not interfere with its binding activity. The best immobilization strategy is one in which the immobilized partner is presented in a conformation and orientation that allows as closely as possible the interaction to proceed as it would in vivo.

Although the structure of a protein may encourage the use of one type of immobilization chemistry in preference to others, the optimal strategy must be empirically determined. Efficient coupling, while important, must not be at the expense of activity. In this section, some immobilization chemistry options for potentially problematic interactants such as acidic proteins, are discussed. Secondly, novel ways to immobilize membrane proteins are presented and finally, some recently developed methods of thiol coupling are addressed.

3.4.1 Immobilization Strategies

3.4.1.1 Amine Coupling

Direct immobilization of proteins using amine coupling is the most commonly used strategy because most proteins contain many potentially reactive primary amine groups (Sect. 3.1.1). This method is cited in more than half of all published papers featuring Biacore systems. Proteins with an isoelectric point (pI) greater than approximately 3.5 can be efficiently preconcentrated close to the sensor surface by electrostatic attraction and immobilized in the presence of a buffer of around pH 5. A typical example of the SPR response during the course of immobilization of a protein via amine coupling is shown in Fig. 13.

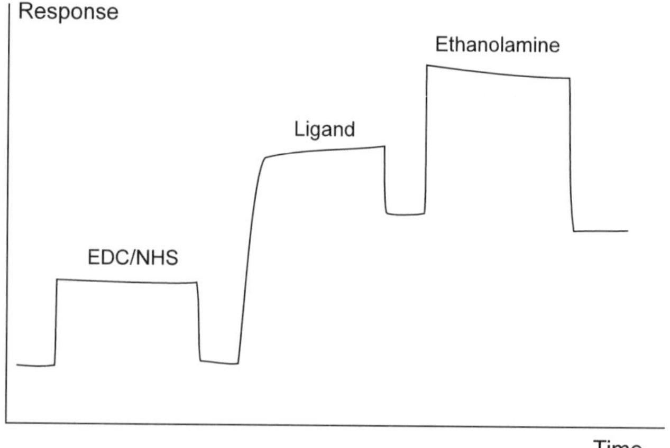

Fig. 13 Sensorgram for a typical amine coupling illustrating the distinction between the amount of protein bound and the amount immobilized

However, at this pH, highly acidic proteins (pI ~ 3 or less) carry a net negative charge and are repelled from the dextran layer. It is not possible to reduce the pH to accommodate the immobilization of highly acidic proteins because the dextran on the chip surface will protonate and become resistant to activation by EDC/NHS. In order to immobilize highly acidic proteins by amine coupling, it is necessary to modify the standard immobilization protocol. Some options are now described.

3.4.1.2
Micelle-Mediated Immobilization of Negatively Charged Proteins

A novel amine coupling method has been developed by Biacore in which an acidic protein is carried by a positively charged micelle (a cluster of oriented surfactant molecules). The micelle–protein complex carries a net positive charge at neutral pH, and is therefore attracted to the sensor surface (Fig. 14).

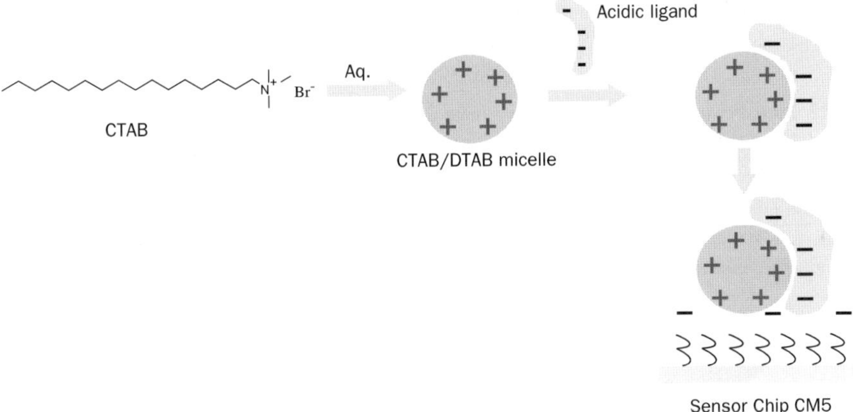

Fig. 14 Hexadecyl-3-methylammonium bromide (CTAB)/dodecyl-3-methylammonium bromide (DTAB) micelle-mediated immobilization of acidic proteins. CTAB/DTAB forms positively charged micelles in aqueous solutions. The micelles bind electrostatically to the negatively charged sensor surface and may thus be used as carriers for acidic proteins

3.4.1.3
Directed Amine Coupling: Protein Modification in Solution

Although immobilization of proteins by amine coupling does not usually inhibit the entropic freedom of macromolecules or significantly change their interaction properties [54], it is nevertheless desirable to present a protein to its binding partner in a directed and uniform orientation. This may be achieved by chemically modifying a specific region of the protein in solution

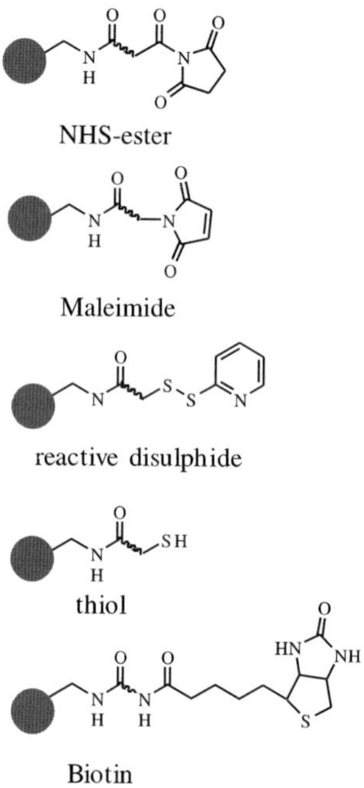

NHS-ester

Maleimide

reactive disulphide

thiol

Biotin

Fig. 15 Modification options. The *wavy lines* illustrate the presence of an undefined chemical structure between the amine group on the protein and the introduced functional group

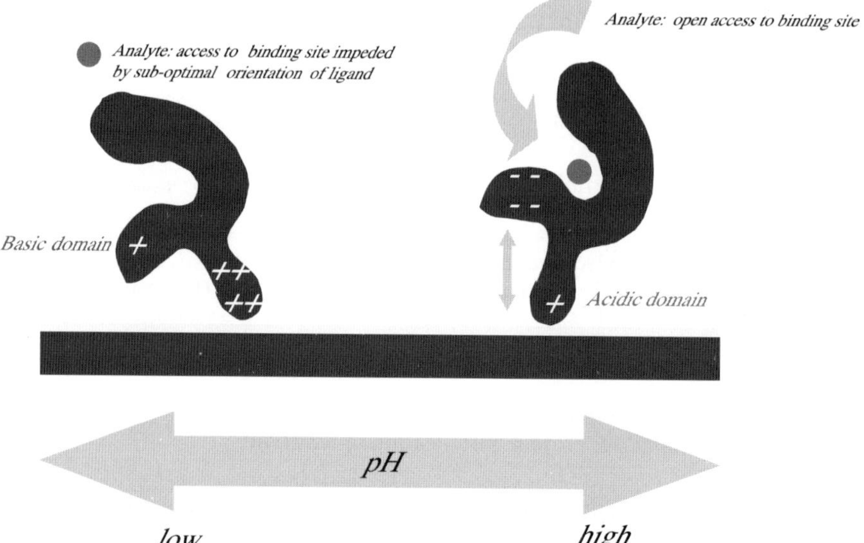

Fig. 16 A protein will have a net positive or negative charge according to the pH of the buffer in which it is dissolved. The distribution of the charges on the protein surface will influence the orientation in which it will approach and immobilize on a charged surface

in order to control the degree of modification or to steer the immobilization process. Some of the most commonly introduced modifications are shown in Fig. 15. Modification in solution may be considered, for example, when handling proteins with high or low pI in order to bring the pI close to the pH at which amine coupling proceeds efficiently.

Coupling at a specific pH may also affect activity, for example, it is possible to predict the orientation in which a protein will immobilize on the chip surface at a particular pH if the spatial distribution of charged amino acids is known (Fig. 16). Alternatively, the protein may be prevented from immobilizing on the surface via sites important to the interaction under study by having analyte present during immobilization, as discussed below [55].

3.4.1.4 Stabilization after Immobilization

Certain proteins are unstable and may deteriorate once they are on the sensor surface, even if the immobilization procedure has worked efficiently. HIV protease, for example, dissociates into its component monomers over time and the gradual reduction in molecular weight on the sensor surface is seen as a baseline drift on the interaction profile. This makes the kinetic analysis of interactions, particularly those involving small molecules, problematic. An extra post-immobilization activation/cross-linking step with EDC/NHS may be considered in order to achieve a stable baseline. The use of cross-linking must be empirically determined for each interaction and the process must not interfere with the activity of the protein.

3.4.1.5 Preservation of Activity During Immobilization

Stabilization during immobilization has been reported, where, for example, the protein kinases, p38α and GSK3β, were immobilized using amine coupling in the presence of a specific reversible inhibitor [56]. This treatment resulted in a more stable sensor surface with much higher specific activity for binding partners (Fig. 17).

3.4.1.6 Protein Stabilization After Capture

Capture protocols may be the most effective way to immobilize proteins, e.g., the protein may contain a molecular tag (Sect. 3.2). Consider, for example, Fig. 18, which depicts a capture system for measuring the binding of ATP or ATP inhibitor to an immobilized kinase.

The kinase in this case was tagged with histidine (His) and was captured on the sensor surface via an anti-His antibody. A capture system does not

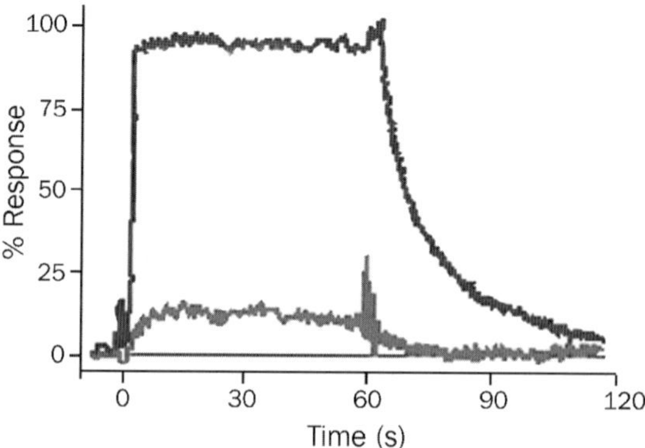

Fig. 17 Comparison of the surface binding capacity of unprotected and protected (stabilized) p38α. Inhibitor (1 μg) was injected over two flow cells containing p38α. One surface contained p38α immobilized using a standard amine coupling procedure (*lower trace*) while the second surface (*upper trace*) contained p38α stabilized by the inhibitor during immobilization

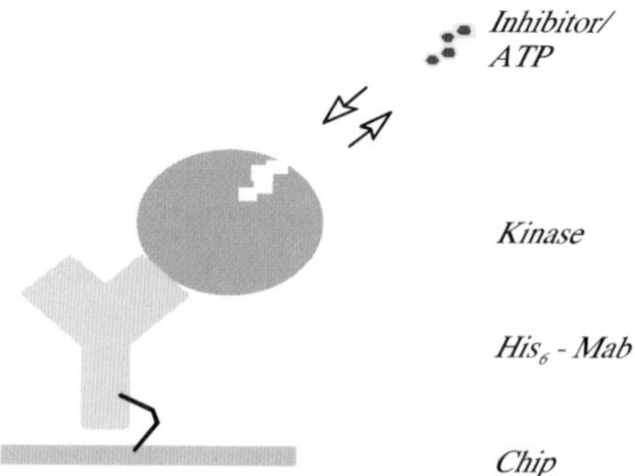

Fig. 18 Capture of a His_6-tagged kinase via an immobilized anti-His antibody

in itself guarantee a stable baseline but, by performing a subsequent cross-linking step, the kinase here was stabilized and remained active, yielding a surface that was open to detailed binding studies of ATP and ATP inhibitors. It is important to realize, however, that the capture surface may not be reused after stabilization.

Large baseline drifts caused by unstable proteins or poor capture may be overcome by using EDC/NHS as a cross-linking step. This step, however, may compromise protein activity if active sites are involved in the cross-linkages. The effect of cross-linking on activity, therefore, must be empirically tested for each system. In general, cross-linking should be as brief as possible; 15 s is often sufficient to achieve acceptable baseline stability, without compromising activity.

3.4.1.7 Immobilizing Membrane Proteins

Membrane spanning proteins are difficult to handle in vitro because they are often functional only when their hydrophobic transmembrane domains are maintained in a native structure. This conservation of structure requires close association with a hydrophobic milieu and presents a practical handling problem in the typical aqueous microenvironment of most in vitro systems.

However, the study of hydrophobic proteins using Biacore systems is possible using protocols specially designed for the purpose. For example, proteins may be firstly incorporated into liposomes and then immobilized on hydrophobic sensor surfaces (Sect. 3.3). It has been reported that lipid bilayers have been tethered to a sensor surface via hydrophilic spacers immobilized on a plain gold chip into which membrane-spanning proteins are then inserted [57]. The emphasis of this technique rests on encasing sensitive protein domains within a lipid microenvironment in which they can assume a native, functional structure.

A recently developed protocol for handling membrane proteins on Biacore's amphiphilic Sensor Chip L1 is *on-surface reconstitution* (OSR). In this process, the membrane protein is firstly solubilized in detergent. Then, it is immobilized on the sensor surface using amine coupling (although OSR may also readily be used in combination with capture). The next step is to inject mixed micelles, which bind to the lipophilic tails on the sensor surface and to the protein. Finally, the detergent is removed, inducing the lipids to form a plasma membrane-like bilayer, linked to the immobilized membrane protein via the natural affinity of lipids for the hydrophobic regions of the proteins (refer to Fig. 12). In this way it is possible to present a uniform, oriented field of plasma membrane proteins within a lipid bilayer on a sensor surface and to study how they interact with binding partners.

3.4.2 Thiol Coupling

Thiol coupling may be considered when using acidic proteins that do not preconcentrate at the chip surface at the pH required for optimal amine coupling, in cases where the protein contains few amino groups or where those present occupy the analyte binding site. Thiol coupling, like amine coupling, may also be considered as a means to control orientation on the sensor surface. Although thiol coupling is an established immobilization procedure, its versatility may be increased by introducing thiol groups to proteins, for example by using the thiolating agent, 2-iminothiolane (Traut's reagent). Here, the sensor surface is prepared for thiol coupling by firstly activating with EDC and NHS followed by PDEA to introduce reactive disulfides. Primary amine groups on the protein are then modified to thiol groups using Traut's reagent. Thiol derivatization can be controlled to produce sensor surfaces with a very high binding capacity.

3.4.2.1
Surface Thiol Coupling

In additional to conventional thiol coupling, the same process may be followed by derivitizing carboxyl groups on the protein (instead of those on the sensor surface) with reactive disulfides. In this way, the protein may then be immobilized to a surface derivatized with thiol groups by treatment with Traut's reagent.

3.4.2.2
Thiol Coupling by Derivatization of Carbohydrate to Maleimide Groups

In this process, protein *cis*-diols or carbohydrates are firstly oxidized to aldehydes using sodium metaperiodate. The aldehyde groups are then modified to thiol-reactive maleimide groups using the bifunctional reagent, *N*-[ε-maleimidocaproic acid] hydrazide (EMCH). Immobilization using standard aldehyde coupling (via the conversion of diols to aldehydes, which then react with hydrazide groups introduced on the sensor surface) is well established. However, thiol coupling may enable the investigator to more tightly control the number of immobilization points between the surface and the immobilized partner. The maleimide-modified protein may then be immobilized on a thiol surface. Thiol coupling to a maleimide surface may also be performed by derivatization of carboxymethyl groups on the sensor surface. In this case, a maleimide surface is generated on the sensor surface, followed by immobilization of a protein possessing thiol groups via a thioether linkage. Proteins immobilized using maleimide coupling are more stable under reducing conditions than those immobilized using thiol coupling via disulfide formation because no thiol-disulfide exchange can take place.

3.4.3
Summary

Alternatives to amine coupling may be considered under a number of circumstances, e.g., when working with acidic proteins or with proteins that may be compromised in their biological activity due to immobilization via primary amino groups located in the analyte recognition site. Membrane spanning proteins may require a hydrophobic microenvironment at the sensor surface in order to remain biologically active and therefore must also be specially

Table 1 Immobilization alternatives to amine coupling

Immobilization protocol	Application examples
Micelle-mediated immobilization	When using highly acidic proteins
Protein modification in solution	To control and favorably orientate proteins on the chip surface for optimal analyte binding
Protein stabilization after direct immobilization or capture	When using unstable proteins, e.g., multimeric or autoproteolytic proteins or an unstable capture system
On-surface reconstitution (OSR) as an alternative to liposome-mediated coupling	When working with membrane-spanning or membrane-embedded proteins that require a hydrophobic microenvironment for their biological activity
Thiol coupling (protein or chip surface)	When using highly acidic proteins, proteins that contain few primary amine groups or where those present may be involved in analyte binding
Maleimide coupling (protein or chip surface)	When thiol coupling is indicated but when the assay is to be run under reducing conditions

handled. Table 1 lists the immobilization alternatives discussed in this section and some of their possible applications.

4 Molecular Recognition Elements

As should be evident from the preceding sections, different types of molecular structures create specific demands to achieve optimal coupling. It is therefore appropriate to describe the specific properties of the respective molecular classes used as interacting partners and to suggest the most suitable immobilization methods for each of them. A more extended general review of the modification and conjugation of different classes of interacting partners may also be found in [25].

4.1 Proteins

Proteins are the most widely used immobilization partners in SPR-based assays but, as a very diverse class of molecules, they are not amenable to a common immobilization strategy. The common denominator is the polypeptide backbone where the individual amino acids supply the functional moieties that can be utilized for immobilization to the sensor surface. The most useful amino acids, together with their corresponding functional groups are: lysine ($-NH_2$), cysteine (– SH), asparagine and glutamine (– COOH), serine and tyrosine (– OH), and histidine (imidazole). The *N*-terminal amino acid residues also constitute a potential linkage moiety via their end amine group. Despite these similarities, the different physical properties related to charge balance and distribution, size, and thermodynamic stability make almost every protein unique with respect to the ease and success of immobilization.

Antibodies are the most homogenous protein class and are also the most frequently used recognition elements for different types of applications. Characterizations of binding properties are important in the selection of therapeutic and diagnostic antibodies and their derivatives. Antibodies are also used for capture of various molecules, or as binders in concentration assays. IgG-type antibodies are composed of an F_c subunit and two Fab′ subunits, constituting in total a molecular weight of 150 kDa [58]. The active antigen-binding regions are localized in the Fab′ subunits and so immobilization to surfaces should ideally be made via the F_c region [59]. Given the size of antibodies this is not normally a problem and they are among the easiest molecules to immobilize. A typical IgG antibody contains 50–70 lysine residues and, when using covalent coupling to an electrophilic functionality, the probability for immobilization via the F_c region is high. There are also alternatives utilizing the carbohydrate residues attached to the F_c region. Mild oxidation of these sugars with periodate generates aldehyde groups, which can react via hydrazide functionalities on the sensor surface. Another way for covalent immobilization involves digestion and reduction of the antibodies into Fab′ fragments, exposing a sulfydryl group from the cysteine residue in the *C*-terminal region. The sulfydryl group is oriented away from the antigen-binding region and can be used for covalent coupling to selective groups such as reactive disulfides and maleimide groups. Certain subclasses of antibodies can also be non-covalently immobilized via the F_c region to protein A or protein G molecules. Antibodies, therefore, constitute a class of proteins that do not normally create any problems in the immobilization step.

Other soluble proteins behave much more heterogeneously than antibodies. The general approach is similar, and covalent immobilization via nucleophilic residues like amine and sulfydryl groups are normally the first method of choice. However, a small protein may be more susceptible to deactivation due to a higher probability that functional groups involved in coupling are close to the interaction site. The most commonly used alternatives involve the introduction of recombinant tags, such as oligo-histidines and other short peptides. The proteins can bind to immobilized antibodies or other capturing agents that specifically recognize these tags. Similarly, recombinant proteins containing larger fusion domains such as GST or Myc can also be generated.

4.2 Peptides

Given their structural similarity to proteins, the principles governing peptide immobilization are comparable. Depending on the composition of the peptide, electrostatic attraction can be utilized in a similar manner as for proteins. Amine- or thiol-based coupling can also be performed under slightly alkaline conditions, using millimolar concentrations of the peptide. Although this straightforward coupling procedure works well in many cases, immobilization of small peptides may need alternative strategies. For example, there is a significant risk that amine groups originating from the lysine residues or the *N*-terminal amino acid are involved in the binding of small peptides to an interaction partner. A thorough analysis of the peptide structure and an evaluation of possible immobilization sites are therefore recommended. Synthetic peptides can be extended with suitable coupling groups. Extra lysines, for example, can be introduced to a region of the peptide that is not involved in the interaction. Spacers that include cysteine residues for use in thiol-based couplings are also favorable alternatives [60]. Another preferred alternative is to biotinylate the peptide in a specific position, optionally followed by chromatographic purification. A well-defined derivative is thereby obtained, which can then be immobilized to streptavidin-modified sensor surfaces.

4.3 Oligonucleotides

Oligonucleotides are composed of negatively charged nucleotide groups that are relatively resistant to covalent coupling under mild aqueous conditions. Although the phosphate ester groups can be used in condensation reactions with nucleophiles such as primary amines, the reaction is slow and water hydrolysis competes unfavorably with the desired reaction. Further, the nucleotide bases are weak nucleophiles and cannot be utilized under conditions normally used for coupling.

The most common alternative for immobilization of oligonucleotides involves the use of biotinylated derivatives. These are conveniently made with standard reagents for oligonucleotide synthesis and biotin can be added at both the 3′ and 5′ ends. Immobilization of the biotinylated oligonucleotide to avidin-modified sensor surfaces is performed in neutral buffer conditions and is normally efficient.

An alternative method has recently been developed that uses 3′ or 5′ amino-derivatized oligonucleotides for coupling to activated carboxymethylated sensor surfaces. Electrostatic repulsion between negative charges on both the surface and the oligonucleotides normally slow this type of reaction. Here, however, the oligonucleotide is mixed with a positively charged detergent like hexadecyltriammonium bromide (CTAB). Under conditions of micelle formation, this leads to a complex that can neutralize the repulsion effects, with the positive micelle acting as a carrier of the oligonucleotide, which is attracted to the negative surface, greatly increasing coupling efficiency [61] (see also Sect. 3.4.1). When applied to carboxymethylated dextran surfaces, the oligonucleotide densities reach levels twice those of streptavidin surfaces.

4.4 Small Organic Molecules

Although applications have been dominated by interactions involving immobilization of proteins, the use of small organic molecules such as hormones, vitamins, and drug candidates with molecular weights typically lower then 700 Da is increasing. Normally, these types of molecules need to be treated differently to those previously described. The type and number of suitable functional groups available for coupling to the sensor surface is unique for each molecule and general immobilization procedures are not applicable. It may even be necessary to synthesize derivatives of the molecule with functional groups in desired positions. This may also be a necessary step in order to minimize interference between the analyte and the immobilized molecule. Also, many organic molecules have very low solubility in aqueous solutions and need to be handled in organic solvents such as DMSO and DMF (optionally in water mixtures) during immobilization. Derivatives of molecules to which groups are introduced to increase water solubility are therefore attractive options. Approaches using electrostatic attraction as described previously are not normally applicable for small molecules.

Molecules with functional groups like aliphatic amines, thiols, aldehydes, or carboxylic groups can normally be covalently linked to suitable corresponding active groups on the sensor surface, as described in Sect. 3.1. Amine coupling is normally performed under aqueous buffer conditions at a concentration between 1 and 50 mM at pH 7–8.5. Thiol coupling proceeds efficiently in near-neutral buffer conditions, while aldehyde condensation to hydrazide-modified sensor surfaces can be performed in a slightly acidic buffer or in the presence of a reducing agent such as cyanoborohydride. Following their activation to reactive esters, molecules with carboxylic groups can likewise be immobilized to sensor surfaces with amines or hydrazide groups. The activation step can either be performed before coupling, utilizing EDC/NHS, or in situ, in the presence of EDC or some other condensation agent.

Small molecules without suitable functional groups need to be derivatized, a procedure normally requiring significant synthetic organic chemistry efforts. This approach is advantageous as the tailoring of the molecular structure creates conditions necessary for a successful outcome of the assay. In particular, the choice of type of functional group and position in the molecule can be optimized in order to avoid interference with other groups or parts of the molecule. The introduction of a spacer is also normally of value, both in order to reduce steric interferences and to increase water solubility. Details of various modification approaches can be found in the literature [25]. The introduction of general tags such as biotin is also applicable to small organic molecules. This route is attractive due to the commercial availability of different derivatives of biotin, optionally with spacer groups.

4.5 Carbohydrates

Interactions that involve carbohydrate structures are important in many biological events, including cell adhesion, apoptosis, and immune responses. Their interactions with proteins are normally weak in affinity and traditional techniques may be difficult to use. SPR detection can therefore be favorably utilized for these studies and the need for such analytical methods is expected to grow within the emerging field of glycomics. However, immobilization of carbohydrates may be a challenge, depending on the explicit nature of the substance.

The primary potential coupling sites for sugars are hydroxyl groups, although there is a high likelihood that the hydroxyls of small sugar compounds (e.g., mono- to oligosaccharides) are important for binding activity. Alternative strategies, similar to those used for other types of small molecules must therefore be applied. The most common approach is to use the anomeric aldehyde group for direct immobilization, or for modification in solution before immobilization. Aldehyde coupling as described in Sect. 3.1.3 is a good alternative in such cases. Alternatively, linker molecules with reactive groups such as thiols can be introduced [62]. The introduction of a biotin derivative is also a highly suitable route.

Polysaccharides can be immobilized in various ways depending on which functional groups are present. Glycosylated proteins can be considered as a special case, as described in Sect. 4.2. The main difference between "pure" polysaccharides and glycoprotein structures is that electrostatically mediated enrichment of the coupling cannot normally be used, depending on the presence of charged groups. For covalent immobilization via nucleophilic or electrophilic groups (e.g., reactive amines, aldehydes, or activated carboxyls) high concentrations of the polysaccharide need to be used and the resulting densities are still relatively modest. A better alternative is to biotinylate a suitable functional group before immobilizing the derivative to streptavidin surfaces, as shown for heparin [63].

5
Spatially Resolved Immobilizations

SPR detection is highly adaptable to multiplexed configurations in miniaturized formats. The flow cells in the original Biacore systems had four measuring spots positioned within a few millimeters (Fig. 19). Prototype systems with eight parallel flow channels have also been described and applied to food analysis applications [64].

In these systems, both the immobilized partner and analyte in solution are delivered to the sensor surface via an integrated microfluidics device. All the steps in the immobilization procedure can therefore be monitored by SPR detection and serve as an important guidance and quality check. The channels for fluid delivery and flow cell structures are made by micromolding of elastomeric materials and are produced with cell widths down to a few hundred micrometers [65]. Spatial distribution of immobilized partners is achieved by addressing the distinct flow cells individually.

The measuring spot density on several subsequently commercialized biosensors has greatly increased, allowing arrays to be probed and generating parallel interaction data. One recent development involves microfluidics systems with hydrodynamic addressing (HA) of the solutions (Fig. 19). By the

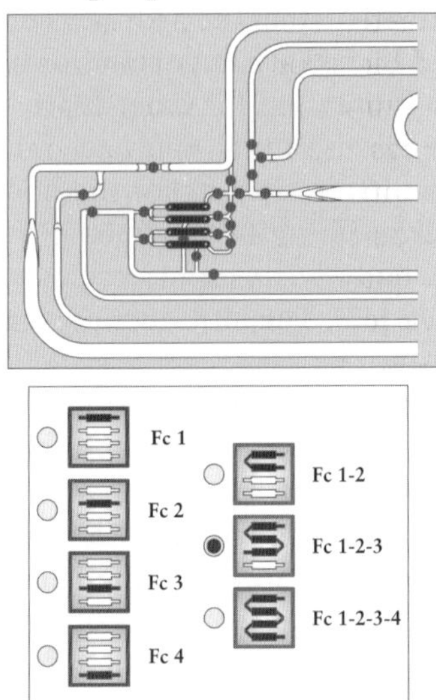

Fig. 19 Configurations and dimensions of various Biacore instrument flow cells. The left-hand flow cell is designed for hydrodynamic addressing

use of two parallel fluid inlets, solutions can be directed over different areas of a sensor surface in a single flow cell and can therefore be used for spatially resolved immobilizations [66]. Besides the obvious advantage of higher sample throughput, this flow cell system allows a more precise and detail-rich analysis to be performed. Firstly, as all interactions are measured simultaneously, highly accurate reference subtraction allows the measurement of very rapid kinetics. Further, by immobilizing several targets in one flow cell, interaction profiles may be directly compared under identical conditions.

In the Biacore A100 system containing four HA flow cells with five immobilization spots per flow cell, assays can be designed either for the maximum number of samples, or to deliver the maximum information per sample. In the first instance, identical immobilizations may be performed in all four flow cells, allowing four different samples to be analyzed in parallel during each analysis cycle, while in the latter, up to 20 different interactants may be immobilized across all four flow cells with one sample per cycle injected in parallel.

What types of application areas may best benefit from a high throughput array system that also delivers high quality kinetic data? As one example, the development of therapeutic mAbs, for example, is a complex and time-consuming process, involving generation, maintenance and above all, screening of thousands of hybridoma clones. Early identification of those hybridomas that produce the best candidate antibodies is a critical step in successful, cost-efficient development. Efficient screening of many hundreds of hybridomas would enable selection of candidates with the best prognoses for clinical success based on their kinetic properties. Secondly, even the most carefully designed and constructed biotherapeutics may be sensed as foreign proteins by the patient, causing an unwanted antibody response. The immunogenicity of newly developed drugs and vaccines is one area that could benefit from sensitive detection of potentially clinically relevant low/medium affinity antibodies, generating data on isotype, subclass specificity, and kinetics from a single system using low quantities of sera.

For larger 2D arrays, various approaches have been employed. Techniques developed during the 1990s, principally for DNA arrays have also been evaluated for SPR detection [67]. Depositions by contact or non-contact means have become the most common alternatives and several commercial arrayers are available. A general review of protein arrays can be found in [68]. One possible limitation with these approaches is that relatively high concentrations need to be employed, due to the requirement of high surface density of active molecules needed for SPR detection. The drop deposition of the solution must therefore be optimized to meet these needs. Ink-jet or piezoelectric printing devices that were originally employed for DNA applications can also be used for proteins, both in aqueous and organic solvents. However, careful optimization is normally needed when these are used for proteins in buffered solutions, as deposits and clogging of the ink-jet heads tend to occur, particular for solutions of high protein concentration. Tendencies for smeared spots and an uneven spread of the density of the immobilized partner across the spot have also been reported.

Another important issue related to deposition of proteins is the tendency for denaturation. When spotting droplets of protein solution in the nanoliter range, evaporation of the solvent will quickly lead to air exposure. Alternative strategies have been employed to minimize evaporation, including printing in high humidity and using 40% glycerol solutions.

Photolithography has become a popular method for DNA arrays, with the Affymetrix approach for site synthesis of DNA probes. Similar approaches can be applied to spatially synthesize combinatorial libraries of peptides and other organic compounds [69]. However, applications within the protein array field are still in a preliminary state. Examples can be found in the literature [70, 71] where photomasks have been used to define the area for immobilization. After exposure to UV light in the presence of an immobilized partner, the surface is washed and the procedure repeated with a new mask and a new preparation of the molecule to be immobilized. The method may be limited by the risks for non-specific binding and the extended time needed to generate large arrays. Although it is tempting to perceive that functional protein arrays will be constructed on a similar scale to DNA arrays, the natural sensitivity of protein structure – essential for function – will almost certainly limit their size; it is simply not possible to control the biological activity of thousands of proteins extracted from their natural microenvironment and immobilized in vitro after lengthy and stringent amplification and purification processes. The functions of proteins are often dependent on domain integrity and demand precisely defined microenvironments. For example, the conformation of transmembrane receptors are difficult to predict in vitro, when the predominantly hydrophobic membrane-spanning domain is dissolved in an aqueous solution, rather than integrated in the protective and stabilizing milieu of the cell membrane. Variables such as immobilization conditions, orientation, and the possibility that the immobilization process may impede or conceal the very binding site of interest are further important considerations. Finally, unwanted adsorption to both the array surface and other proteins in a complex mixture may complicate the interpretation of results from a multiplexed array.

6 Outlook

The development of tools for the immobilization of molecules to sensor surfaces has been dramatic over the last two decades. This has been driven by the availability of better surface analysis techniques, including the commercialization of sensitive and reliable systems, as well as better tools for manipulation of proteins and other biomolecules. Growing interest in understanding biomolecular interactions on a quantitative level have also driven this development. These trends will almost certainly continue to drive the field of immobilization. The availability of large protein libraries for proteomics applications is increasing, which will drive the need for larger scale protein interaction analysis. This leads to a demand for higher throughput systems and more parallel and streamlined immobilization strategies. Better tools and techniques for the creation of 2D arrays will likely emerge, with the twin objectives of improving control over immobilization levels and retaining the activity of the protein.

Another field of development relates to improved tags for surface coupling and other conjugation steps. Fusion tags have great potential as immobilization sites to surfaces and can be used for general parallel coupling methods of arrays. However, several of the presently available tags are based on protein or peptide structures that significantly increase the size, which may cause expression problems during the protein fusion step. An additional consequence here is that the tag size may affect the activity of the protein. The limited stability in the bond between the tag and the capturing molecule can also be a problem. Consequently, it can be hoped that smaller tags with better binding properties to their capture partners will be developed in the coming years.

Considering the importance of membrane-associated proteins (particularly GPCRs) as drug targets, the limited range of options for sensor surface immobilization will certainly stimulate continued efforts to find better methods. Increases in the attainable surface densities of membrane proteins with sustained drug binding activity are necessary for their interrogation using label-free SPR detection methods. Some promising advances have been described during recent years [50, 52, 53] and these methods are likely to be further improved. The techniques to express, purify, and reconstitute membrane-associated proteins will also advance, increasing the options for their immobilization.

References

1. Liedberg B, Nylander C, Lundström I (1983) Sensor Actuator 4:299
2. Lin JN, Chang IN, Andrade JD, Herron JN, Christensen DA (1991) J Chromatogr 542:41
3. Nuzzo RG, Allara J (1983) JvAm Chem Soc 105:4481
4. Ulman A (ed) (1998) Thin films: self-assembled monolayers of thiols. Academic, San Diego
5. Löfås S, Johnsson B (1990) J Chem Soc Chem Commun 21:1526
6. Bergström J, Johnsson B, Löfås S (1990) Sensing surfaces capable of selective biomolecular interactions, to be used in biosensor systems. Patent appl. WO90/05303
7. Bishop AR, Nuzzo RG (1996) Curr Opin Colloid Interface Sci 1:127
8. Ostuni E, Chapman RG, Holmlin RE, Takayama S, Whitesides GM (2001) Langmuir 17:5605

9. Sigal GB, Mrksich M, Whitesides GM (1998) J Am Chem Soc 120:3464
10. Löfås S (1995) Pure Appl Chem 67:829
11. Löfås S, Malmqvist M, Rönnberg I, Stenberg E, Liedberg B, Lundström I (1991) Sensor Actuator B Chem 5:79
12. Lundström I (1994) Biosens Bioelectron 9:725
13. Day YSN, Baird CL, Rich RL, Myszka DG (2002) Protein Sci 11:1017
14. Disley DM, Blyth J, Cullen DC, You HX, Eapen S, Lowe CR (1998) Biosens Bioelectron 13:383
15. Frey BL, Corn RM (1996) Anal Chem 68:3187
16. Daniels PB, Deacon JK, Eddowes MJ, Pedley DG (1988) Sensor Actuator 15:11
17. Spinke J, Liley M, Guder HJ, Angermaier L, Knoll W (1993) Langmuir 9:1821
18. Mecklenburg M, Danielsson B, Winqvist F (1997) Broad specificity affinity arrays: a qualitative approach to complex sample discrimination. Patent appl. PCT/EP 97/03317
19. Alon R, Bayer EA, Wilchek M (1992) Eur J Cell Biol 58:271
20. Prime KL, Whitesides GM (1991) Science 252:1164
21. Lahiri J, Isaacs L, Tien J, Whitesides GM (1999) Anal Chem 71:777
22. Harder P, Grunze M, Dahint R, Whitesides GM, Laibinis PE (1998) J Phys Chem B 102:426
23. Chapman RG, Ostuni E, Takayama S, Holmlin RE, Yan L, Whitesides GM (2000) J Am Chem Soc 122:8303
24. Hermanson GT, Mallia AK, Smith PK (1992) Immobilized affinity ligand techniques. Academic, New York
25. Hermanson GT (1995) Bioconjugate techniques. Academic, New York
26. Myszka DG (1999) J Mol Recognit 12:390
27. Johnsson B, Löfås S, Lindquist G (1991) Anal Biochem 198:268
28. Stenberg E, Persson B, Roos H, Urbaniczky C (1991) J Colloid Interface Sci 143:513
29. Löfås S, Johnsson B, Tegendal K, Rönnberg I (1993) Colloids Surf B: Biointerfaces 1:83
30. Matson RS, Little MC (1988) J Chromatogr 458:67
31. Sprik M, Delamarche E, Michel B, Rothlisberger U, Klein ML, Wolf H, Ringsdorf H (1994) Langmuir 10:4116
32. Johnsson B, Löfås S, Lindquist G, Edström Å, Müller Hillgren RM, Hansson A (1995) J Mol Recognit 8:125
33. Stuchbury T, Shipton M, Norris R, Malthouse JPG, Brocklehurst K, Herbert JAL, Suschitzky H (1975) Biochem J 151:417
34. O'Shannessy DJ, Brigham-Burke M, Peck K (1992) Anal Biochem 205:132
35. Hoffman WL, O'Shannessy DJ (1988) J Immunol Methods 112:113
36. Gershon PD, Khilko S (1995) J Immunol Methods 183:65
37. Nieba L, Nieba-Axmann SE, Persson A, Hämäläinen M, Edebratt F, Hansson A, Lidholm J, Magnusson K, Frostell-Karlsson Å, Plückthun A (1997) Anal Biochem 252:217
38. Sigal GB, Bamdad C, Barberis A, Strominger J, Whitesides GM (1996) Anal Chem 68:490
39. Radler U, Mack J, Persike N, Jung G, Tampé R (2000) Biophys J 79:3144
40. Inagawa J, Inomota N, Suwa Y (2004) Immobilization method. Patent appl. WO 2004/046724 A1
41. Tripet B, De Crescenzo G, Grothe S, O'Connor-McCourt M, Hodges RS (2002) J Mol Biol 323:345
42. Kindermann M, George N, Johnsson N, Johnsson K (2003) J Am Chem Soc 125:7810
43. Kalb E, Frey S, Tamm L (1992) Biochem Biophys Acta 1103:307

44. Ohlsson PÅ, Tjärnhage T, Herbai E, Löfås S, Puu G (1995) Bioelectrochem Bioenerg 38:137
45. Cooper MA, Try AC, Carroll J, Ellar DJ, Williams DH (1998) Biochem Biophys Acta 1373:101
46. Lang H, Duschl C, Vogel H (1994) Langmuir 10:197
47. Spinke J, Yang J, Wolf H, Liley M, Ringsdorf H, Knoll W (1992) Biophys J 63:1667
48. Cooper MA, Hansson A, Löfås S, Williams DH (2000) Anal Biochem 277:196
49. Erb EM, Chen X, Allen S, Roberts CJ, Tendler SJB, Davies MC, Forsén S (2000) Anal Biochem 280:29
50. Karlsson OP, Löfås S (2002) Anal Biochem 300:132
51. Stora T, Dienes Z, Vogel H, Duschl C (2000) Langmuir 16:5471
52. Yoshima-Ishii C, Boxer SG (2003) J Am Chem Soc 125:3696
53. Svedhem S, Pfeiffer I, Larsson C, Wingren C, Borrebaeck C, Höök F (2003) ChemBioChem 4:339
54. Myszka DG, Abdiche YN, Arisaka F, Byron O, Eistenstein E, Hensley P, Thomson JA, Lombardo CR, Schwarz F, Stafford W, Doyle ML (2003) J Biomol Tech 14:199
55. Casper D, Bukhtiyarova M, Springman EB (2003) Biacore J 3:4
56. Casper D, Bukhtiyarova M, Springman EB (2004) Anal Biochem 325:126
57. Sevin-Landais A, Rigler P, Tzartos S, Hucho F, Hovius R, Vogel H (2000) Biophys Chem 85:141
58. Killard AJ, Deasy B, O'Kennedy R, Smyth MR (1995) Trends Anal Chem 14:257
59. Lu B, Smyth MR, O'Kennedy R (1996) Analyst 121:29R
60. Khilko SN, Corr IN, Boyd LF, Lees A, Inman JK, Margulies DH (1993) J Biol Chem 268:15425
61. Karlsson R, Sjödin A (2004) Immobilization method and kit therefore. Patent appl. US 2004/0241724 A1
62. Smith EA, Thomas WD, Kiessling LL, Corn RM (2003) J Am Chem Soc 125:6140
63. Caldwell EE, Andreasen AM, Blietz MA, Serrahn JN, Vandernoot V, Park Y, Yu G, Linhardt RJ, Weiler JM (1999) Arch Biochem Biophys 361:215
64. Situ C, Crooks SRH, Baxter AG, Ferguson J, Elliot CT (2002) Anal Chim Acta 473:143
65. Sjölander S, Urbaniczky C (1991) Anal Chem 63:2338
66. Myszka DG (2004) Anal Biochem 329:316
67. Schumakaer-Perry JS, Zareie MH, Aebersold R (2004) Anal Chem 76:918
68. Mann CJ, Stephens SK, Burke JF (2004) Production of protein microarrays. In: Kambhampati D (ed) Protein microarray technology. Wiley, Weinheim
69. Jacobs JW, Fodor SPA (1994) Trends Biotechnol 12:19
70. Rozsnyai LF, Benson DR, Fodor SPA, Schultz PG (1992) Angew Chem Int Ed Engl 31:759-761
71. Delamarche E, Sundarababu G, Biebuyck H, Michel B, Gerber C, Sigrist H, Wolf H, Ringsdorf H, Xanthopoulos N, Mathieu HJ (1996) Langmuir 12:1997

Part III
Applications of SPR Biosensors

Springer Ser Chem Sens Biosens (2006) 4: 155–176
DOI 10.1007/5346_018

Published online: 8 June 2006

Investigating Biomolecular Interactions and Binding Properties Using SPR Biosensors

Iva Navratilova · David G. Myszka (✉)

Department of Biochemistry, University of Utah, Salt Lake City, Utah USA
dmyszka@cores.utah.edu

Keywords Biomolecules · Kinetics · Surface plasmon resonance

1 Introduction

SPR-based biosensors can measure the interactions of biomolecules directly without the need for labeling. This feature has allowed these analytical instruments to become essential tools for characterizing molecular interactions. The ability to directly measure interactions in real time allows us to quantitatively determine kinetic parameters, thermodynamics, and concentration, or qualitatively characterize relationships between ligands and analytes. Due to the fast response and high sensitivity of SPR-based biosensors compared to other technologies such as enzyme or radiolabeling methods, biosensors can be used to study a large variety of biomolecular mechanisms, ranging from protein–protein, antibody–antigen, and receptor–ligand interactions to the characterization of even low molecular weight compounds. Progress in surface chemistry enables the use of SPR-based platforms to facilitate capture of hydrophobic compounds such as lipids to study membrane-associated receptors. Higher-throughput SPR biosensors with parallel sample delivery or array-based SPR systems are expanding the technology's applications. In this chapter we will review the main applications of SPR-based biosensors using examples from a variety of biological systems. We will also review some of the key requirements in properly analyzing biosensor data, including processing and fitting methods.

2 Data Processing

Data obtained from biosensors are usually affected by the position on the resonance unit scale, noise, non-specific responses and other artifacts that complicate further presentation. Therefore raw data need to be processed to ensure their comparability. Although many different SPR-based platforms have been developed and data processing can differ slightly, the basic procedures described in this section can be applied with minor adjustments to most of them. Given the importance of data processing, we take time here to describe the general procedures before delving into specific biosensor applications.

The most common commercially available SPR-based biosensors are Biacore 2000 and 3000 systems, which are equipped with a four flow-cell fluidic system. One flow-cell is used as a reference to subtract possible non-specific signal and correct for refractive index changes, injection noise, and instrument drift. Several software packages are available to simplify data analysis for Biacore experiments such as Scrubber (Biologic Software, Australia) or BiaEvaluation (Biacore AB, Uppsala, Sweden). Raw data obtained from Biacore instruments are usually spread according to the actual SPR response in a wide range of response units (RU) for different flow-cells (Fig. 1a). The first step in data processing is to zero the response just before the analyte injection. This can be performed by subtracting an average of the response in a small interval just prior to the start of the injection. The second step is to align the responses so that all injections start at the same point. Typically this can be done by keying on the bulk refractive index jump that often occurs as sample is introduced. Double referencing is a process we introduced a number of years ago that can significantly improve the quality of the data, particularly when working at low response levels. In the first step, signal collected from the reference flow-cell is subtracted from the data obtained for reaction surfaces. The second step is to subtract an average of

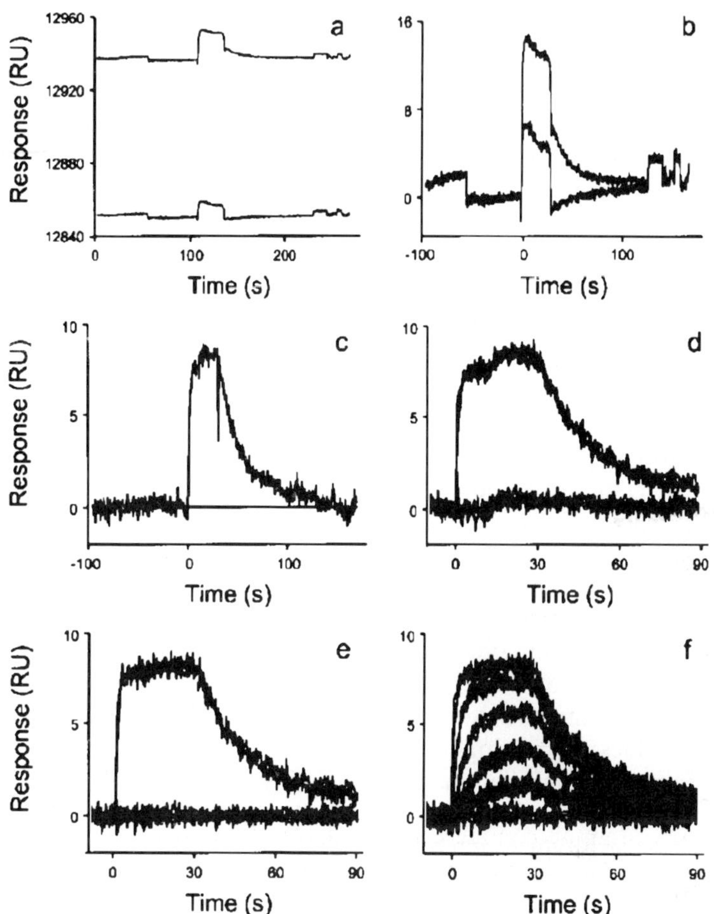

Fig. 1 Processing biosensor data. **a** Raw data from the biosensor for 233 nM IL-2 injected over a receptor surface (*top trace*) and reference surface (*bottom trace*). **b** Data sets were zeroed on the y and x axis just prior to the start of the injection. **c** Data from the reference surface was subtracted from the data from the reaction surface. **d** Overlay of four replicate injections of 233 nM IL-2, as well as running buffer blank. **e** Responses after subtracting the average of the blank injections from both the sample and blank data. **f** Overlay of a series of IL-2 injections (233, 78, 26, 8.6, 2.9, and 0 nM) replicated four times each. Reproduced from [1] with permission from John Wiley and Sons © 1999

the responses obtained for a set of buffer injections. In Fig. 1 we present all the steps required to process data for a protein–receptor interaction, in this case interleukin-2 binding to the alpha receptor subunit. Panel F shows the resulting processed data for a concentration series of IL-2 (0–233 nM) performed in four replicates for each concentration [1]. This is an example of high quality data which is now suitable for more advanced curve fitting including kinetic analysis.

3 Kinetics of Biomolecular Interactions

The number one advantage of optical SPR biosensors is their ability to measure complex formation in real time. This makes it possible to obtain quantitative information about binding interactions including the assembly and break down process. The majority of binding interactions that we encounter on a routine basis are simple bimolecular interactions. Two molecules must come together in space to form a complex. We typically depict these systems as a simple A + B goes to AB reaction as shown below:

$$\mathrm{A} + \mathrm{B} \underset{k_\mathrm{d}}{\overset{k_\mathrm{a}}{\rightleftharpoons}} \mathrm{AB}\,.$$

The rates of complex formation (k_a) and breakdown (k_d) are governed by the intrinsic association and dissociation rates, respectively. This kinetic information reflects the binding mechanism of molecules, which can provide detailed insights into structure and function.

We commonly employ global analysis of interaction data in order to extract accurate estimates of the binding constants. Global analysis means that all the responses within a data set are fit simultaneously using the same set of rate constants. Global analysis of a wide range of analyte concentrations provides a robust method to discriminate between different reaction models. However, it does require high-quality data. Fortunately with improvements in experimental design and data processing, as discussed above, it is fairly routine to collect SPR data that is of high-enough quality for global fitting. Nevertheless, each type of interaction needs to be optimized for experimental conditions, involving density of immobilized ligand, flow rate, and analyte concentrations. A simple one-to-one model is not always sufficient to fit interactions influenced by mass transport and this event must be included for systems where the rate of analyte binding to immobilized ligand is equal or faster than the diffusion of analyte to the ligand surface [2, 3]. The mass transport coefficient k_m then characterizes the rate of analyte A_0 diffusion to and from the reaction surface:

$$\mathrm{A}_0 \underset{k_\mathrm{m}}{\overset{k_\mathrm{m}}{\rightleftharpoons}} \mathrm{A} + \mathrm{B} \underset{k_\mathrm{d}}{\overset{k_\mathrm{a}}{\rightleftharpoons}} \mathrm{AB}\,.$$

The effect of mass transport can be also observed when comparing the same interaction at different ligand surface densities and flow rates [4, 5]. Generally speaking, when the goal of the analysis is to determine binding kinetics, it is often best to work with low ligand density surfaces and high flow rates to minimize these effects.

Software packages such as Scrubber, Clamp or BiaEvaluation were developed to help analyze processed data to obtain kinetic parameters for interactions. Generally, fitting procedures are based on subtracting the simulated data from the experimental data in order to calculate the chi-squared value (χ^2). The initial estimates for the floated parameters are then adjusted automatically to minimize χ^2 using common minimization algorithms [6].

An example of a global fit including the mass transport step for a protein–antibody interaction at different antibody surface densities is shown in Fig. 2a [7]. In this case the apparent binding rate increases as the density of the monoclonal antibody decreases. A simple one-to-one model was not sufficient to fit those interactions and a mass transport step had to be incorporated. An example of another global fit of kinetic data is shown in Fig. 2b. In this example the authors used Biacore instruments to study binding of ankyrin repeat proteins to different targets: maltose binding protein and eukaryotic kinases JNK2 and p38 [8]. A final example of global fitting is shown in Fig. 2c. In this example the authors used kinetic analysis to determine differences between *N*-terminal RNA-binding domains of nucleolin (RBD12) bound to natural pre-rRNA target (b2NRE) and in vitro selected target sNRE [9]. Together, these examples illustrate that it is possible to globally fit data from a variety of systems. Because global analysis is such a stringent test of a reaction mechanism these results further validate the biosensor technology as a biophysical research tool.

4 Equilibrium Analysis

Equilibrium data can be determined from SPR data that have reached steady response levels during the association phase (Eq. 1) or can be calculated from the ratio of the association and dissociation constants (Eq. 2) determined from kinetic analysis:

$$\mathrm{AB_{eq}} = \mathrm{AB_{max}}\left(1/(1 + K_\mathrm{D}/[\mathrm{A}])\right) \tag{1}$$

$$K_\mathrm{D} = k_\mathrm{d}/k_\mathrm{a} \tag{2}$$

$\mathrm{AB_{eq}}$ represents the average of the response signal at equilibrium in defined interval for each concentration of analyte [A]. $\mathrm{AB_{max}}$ is the maximum response in RU that can be obtained for analyte binding depending on the number of binding sites available on the surface. K_D is then calculated by non-linear least squares fit to the data obtained from Eq. 1. Equilibrium analysis represents a simple and fast way of analyzing data to obtain affinity constants for interactions that rapidly reach equilibrium and is efficient as an accompanying method for kinetic data analysis. In the following example (Fig. 3), the authors used both equilibrium analysis and global fitting to determine kinetic constants and affinity of interaction between immobilized peptides pY2267 and pY2327 (representing binding sites on Ros receptor tyrosine kinase) and a GST-fusion of the *N*-terminal SH2 domain of SHP-1 protein tyrosine phos-

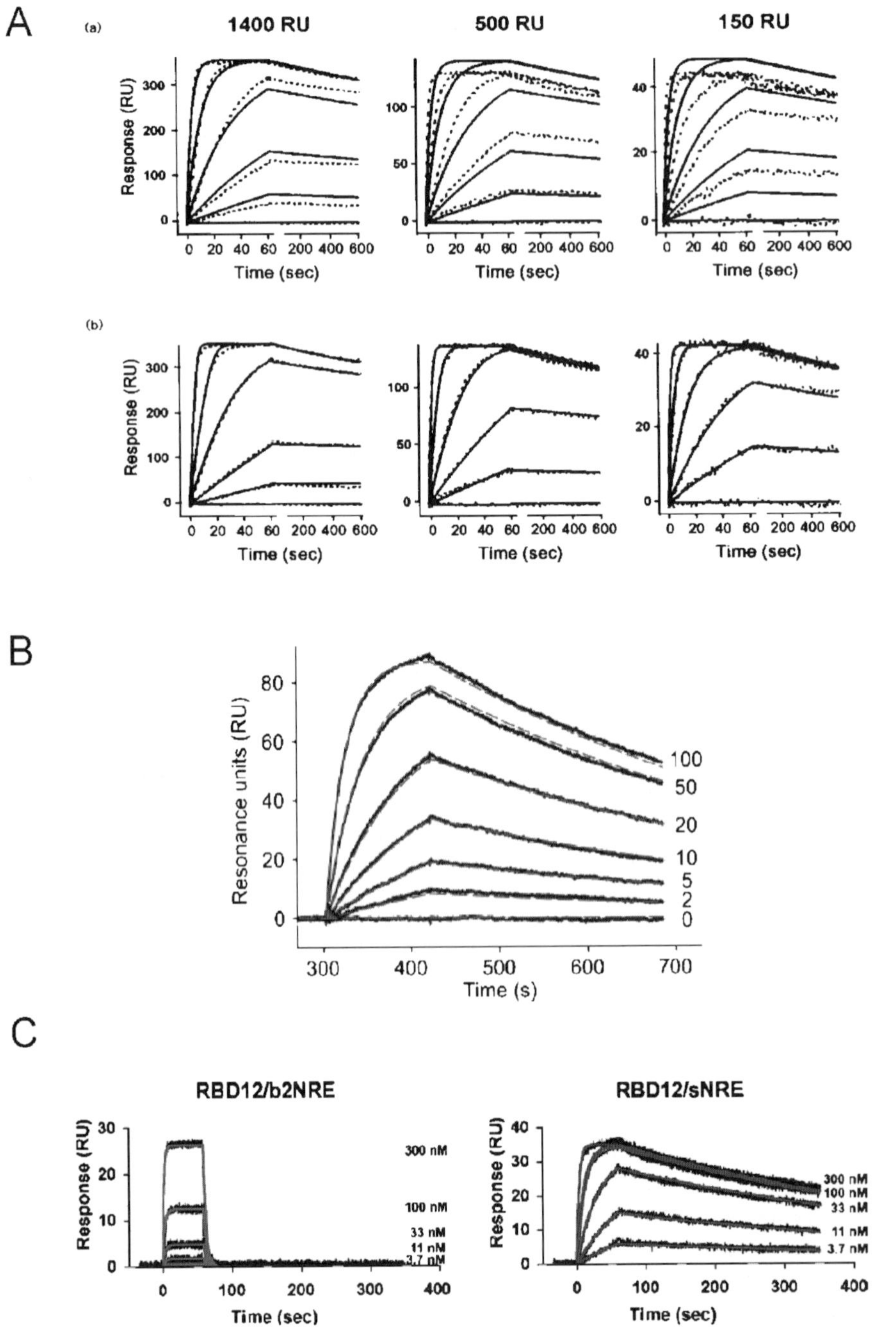

A

B

C

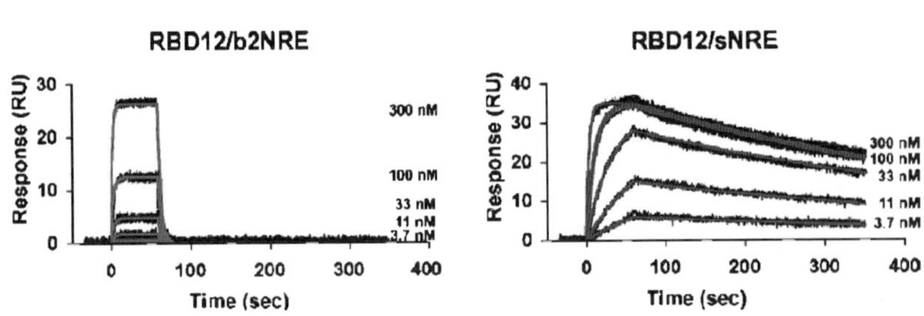
RBD12/b2NRE
Response (RU)
Time (sec)
300 nM
100 nM
33 nM
11 nM
3.7 nM
RBD12/sNRE
Response (RU)
Time (sec)
300 nM
100 nM
33 nM
11 nM
3.7 nM

Fig. 2 ◀ **a** Global analysis of a protein–antibody interaction. The response data (*dashed lines*) are shown for a series of protein concentrations (0, 8.3, 25, 75, 225, and 675 nM) injected over three different surface densities of captured monoclonal antibody (MoAb) (1400, 500, and 150 RU). *a* Best fit to a simple bimolecular interaction model (*solid lines*). *b* Best fit to a two-step mass transport-limited bimolecular interaction model (*solid lines*). The standard deviation of residuals for plots *a* and *b* were 35 and 2.2 RU, respectively. The best fit values of the parameters in *b* were $k_a = 1.2 \times 10^6 \pm 2 \times 10^4\ M^{-1}\ s^{-1}$, $k_d = 2.9 \times 10^{-4} \pm 7 \times 10^{-6}\ s^{-1}$, $k_m = 6.4 \times 10^{-6} \pm 7 \times 10^{-8}\ M^{-1}\ s^{-1}$, $B_{max\,1} = 357 \pm 0.5$ RU, $B_{max\,2} = 132 \pm 0.4$ RU, and $B_{max\,3} = 43 \pm 0.3$ RU. Reproduced from [7] with permission from Elsevier © 1997. **b** Biacore analysis of off7. Different concentrations of off7 (0, 2, 5, 10, 20, 50, and 100 nM) were applied to a flow-cell with immobilized MBP for 2 min, followed by washing with buffer. The global fit is indicated in the figure by *dashed lines*. Reproduced from [8] with permission from the original authors and Nature Publishing Group © 2004. **c** Comparison of the kinetics of RBD12/b2NRE and RBD12/sNRE interactions. Representative sensorgrams are shown for Biacore analyses of the interaction of injected RBD12 protein with biotinylated target RNAs coated on streptavidin sensor chips. *Black lines* represent the binding responses for three random-order replicate injections of protein at 3.8, 11, 33, 10, and 300 nM over the RNA surfaces. Protein was injected at time zero and exposed to the surface for 60 s (association phase), followed by 300 s flow of running buffer during which dissociation could be observed. Note the rapid dissociation of the RBD12/b2NRE complex (*left panel*) compared to the RBD12/b2NRE complex (*right panel*). Each data set was fit with the global analysis program CLAMP, using a single 1 : 1 interaction model. The fitted model is indicated by the *gray lines* on *top* of the data. Reproduced from [9] with permission from Elsevier © 2004

phatase [10]. Using an equilibrium assay it was found that SHP1-N-SH2 binds to a phosphopeptide representing the Ros pY2267 site with a K_D of 217 nM. Ros pY2327 situated at the *C*-terminus of Ros represents another binding site for SHP-1 with a K_D of 171 nM.

5 Thermodynamics

The ability to collect binding data at different temperatures makes it possible to determine thermodynamic properties using SPR. Transition state theory relates the rate constant of a reaction to an equilibrium constant between the reactants and the transient state. The scheme of relationship of free energy and reaction state is represented in Fig. 4. Activation energy is required during the association process to form the transition state. The more energy required, the slower the association rate. Experiments can be performed by measuring kinetic parameters at different temperatures for an interaction, typically from 4 to 40 °C.

In the following example the authors studied the thermodynamics of several HIV-protease inhibitors (Amperavir, Indinavir, Lopinavir, Nelfinavir, Ritonavir, Saquinavir) interacting with the immobilized protein [11]. In Fig. 5

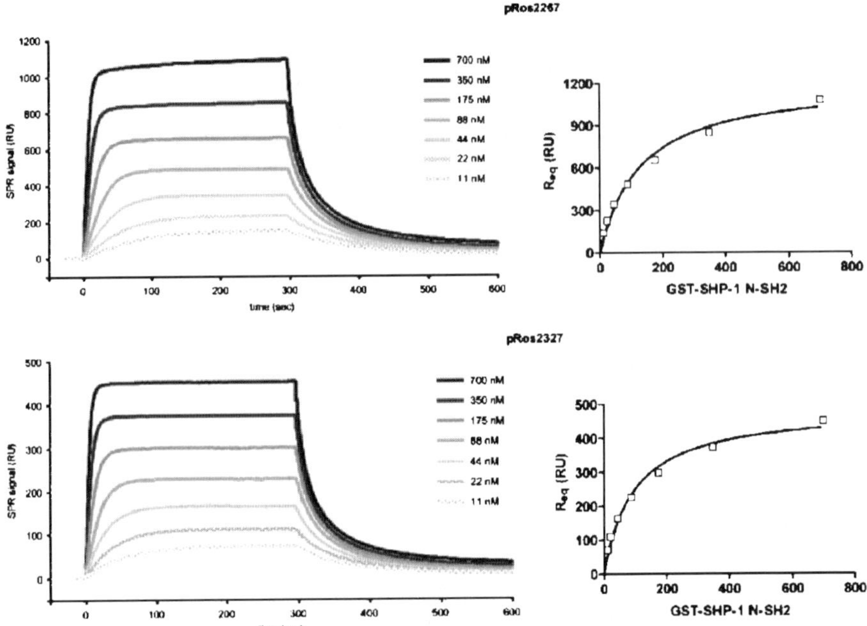

Fig. 3 Synthetic phosphopeptides representing sequences around Ros pY2267 and pY2327 were analyzed for binding of the *N*-terminal SH2 domain of SHP-1 by surface plasmon resonance. Representative experiments are shown (*left panel*). The data were fitted to determine the kinetic constants k_a and k_d. Signals at equilibrium were also fit using an equilibrium analysis to determine the affinities of the interactions (*right panels*). Reproduced from [10] with permission from the Company of Biologists © 2004

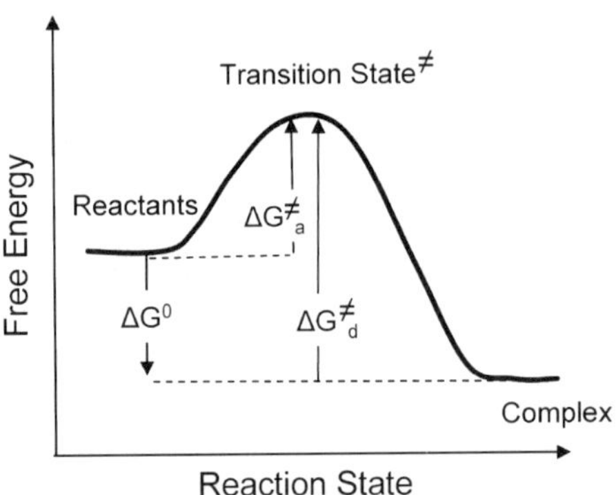

Fig. 4 Free energy profile for interaction. $\Delta G_a^{\neq}$ and $\Delta G_d^{\neq}$ are the changes in Gibbs free energy required for the formation of the transition state starting from reactants and complex, respectively. ΔG^0 is the change in Gibbs free energy between reactants and complex

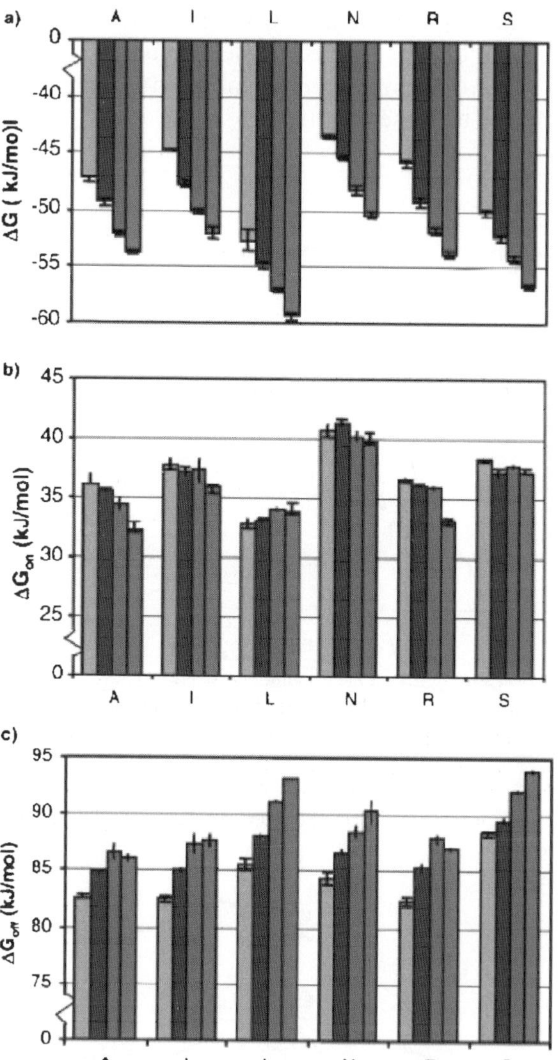

Fig. 5 Temperature dependence of the change in free energies. *Light blue* 5 °C; *dark blue* 15 °C; *magenta* 25 °C; and *red* 35 °C. **a** Equilibrium (ΔG); **b** association (ΔG_{on}); and **c** dissociation (ΔG_{off}). The *y*-axis is adjusted to show a $\sim 20\,\text{kJ}\,\text{mol}^{-1}$ range to facilitate comparison of the different inhibitors. Reproduced from [11] with permission from John Wiley and Sons © 2004

there are three plots representing equilibrium ΔG (a), association ΔG_{on} (b), and dissociation ΔG_{off} (c) energy for each inhibitor bound at 5, 15, and 25 °C. The change in ΔG calculated from the affinity at each temperature provides a measure of stability of complex. The most stable protease-inhibitor complex was formed by Lopinavir (ΔG ranged from -52 to $-59\,\text{kJ mol}^{-1}$). The least stable complex was formed by Nelfinavir (ΔG ranged from -43 to $-50\,\text{kJ mol}^{-1}$). The ΔG became more negative with increasing temperature, corresponding to the increase of affinity.

6 Qualitative Analysis

Along with quantitative kinetic analysis, biosensors can be used to obtain qualitative information for specific processes. By immobilizing the ligand on a surface, a number of different analytes can be studied for binding to the

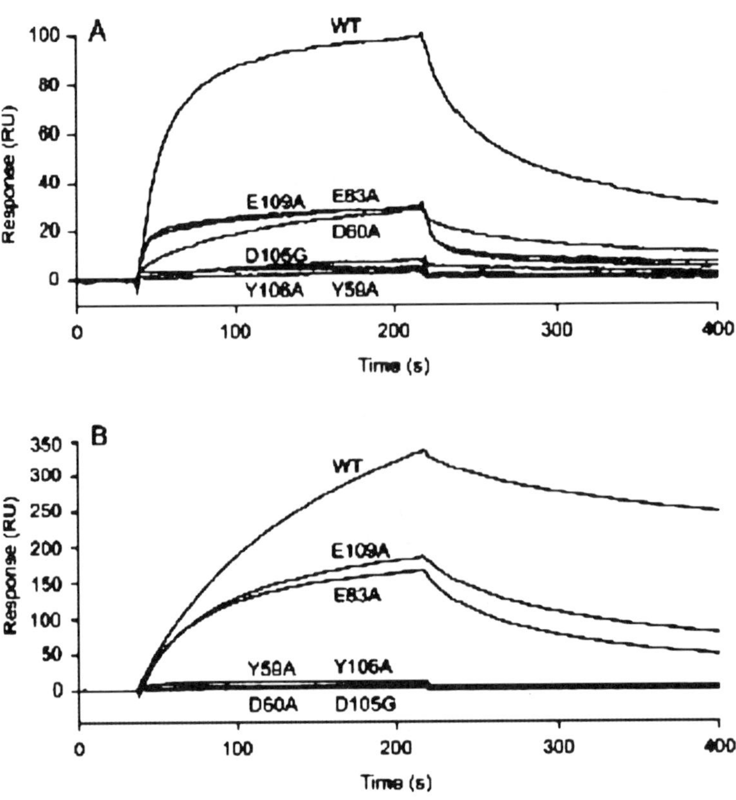

Fig. 6 Analysis by surface plasmon resonance spectroscopy of the interaction between Map19 variants and immobilized MBL or L-ficolin. **a** Interaction with MBL. **b** Interaction with L-ficolin. MBL and L-ficolin were immobilized on the sensor chip. Wild-type Map19 and its Y59A, D60A, D105G, Y106A, and E109A mutants were each injected at a concentration of 50 nM. Reproduced from [12] with permission from the American Society for Biochemistry and Microbiology © 2004

ligand. This method of analysis is convenient when a variety of analytes are being characterized. It helps to determine which compounds bind to immobilized ligand and also which concentrations of analyte are suitable for analysis. In the following example (Fig. 6) the interactions between variants of MAp19 and immobilized mannan-binding lectin (MBL) or L-ficolin [12] were measured. According to the obtained results mutation of Glu^{83} to Ala significantly increased the K_D values for both MBL and L-ficolin. Mutations D60A, D105G, Y59A, Y106A, and E109A decreased the ability of MAp19 to associate with MBL and L-ficolin. The Y59A, D105G, and Y106A mutations virtually abolished interaction with either protein. Mutations E109A and D60A abolished binding to L-ficolin and increased the K_D value for MBL.

7 Competition Analysis

Competition analysis can overcome problems related to interactions that are difficult to analyze directly. Among these types of interactions are systems where the analyte has too low a molecular weight and the sensitivity of the biosensor is not high enough to detect its binding to the immobilized ligand. Another example can be interactions where it is hard to design an assay in a certain way to measure analyte binding because the ligand surface cannot be sufficiently regenerated. There are different mechanisms that can be applied for competition analysis. One of them is based on competition of analyte and competitor with different molecular weights for binding to the ligand surface. In this case, the competitor is typically a compound with lower molecular mass. When binding to the immobilized ligand the response decreases in the presence of increasing concentration of analyte (Fig. 7, left panel). A special kind of competition analysis is analyte inhibition analysis and involves immobilizing the ligand and incubating the analyte with inhibitor at different concentrations for a defined time period. When the inhibitor binds to the analyte it inactivates it for binding to the ligand surface (Fig. 7, right panel). The higher the concentration of inhibitor present in the sample, the

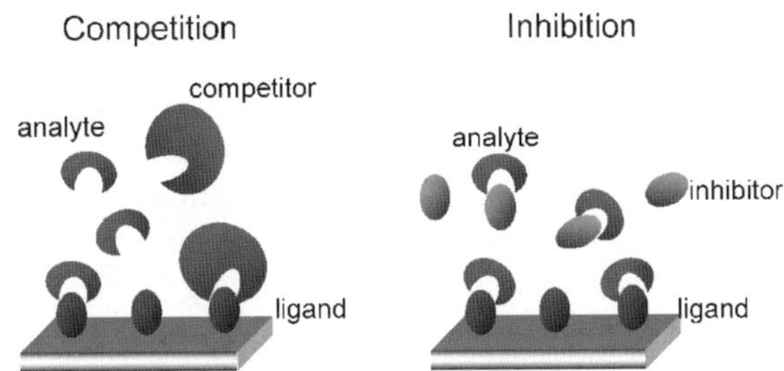

Fig. 7 Scheme for competition reactions

Fig. 8 Peptide 12p1 inhibition of binding of gp120 to CD4 and 17b. CD4 (**a**) and 17b (**b**) ▶ were immobilized on a CM5 sensor chip in a BIA3000 instrument. YU2 gp120 (50 nM) was passed over each surface in the absence (0 nM) or presence of 821 nM to 6.57 μM 12p1. Buffer injections and control surface binding have been subtracted to obtain all curves. Experiments were repeated twice in duplicate with similar results. Data from one experiment are shown. **c** Log plot for determining IC_{50} for 12p1 inhibition of binding to CD4 (♦) and 17b (▲). Curves were fit using SigmaPlot, and the 12p1 concentration at which the initial rate of gp120 binding was half of that without peptide was designated the IC50. Reproduced from [13] with permission from the American Chemical Society © 2004

lower the amount of competitor available for binding to the ligand, and the lower the response observed.

An example of analyte inhibition analysis is demonstrated in Fig. 8. These authors studied peptide 12p1 inhibition of gp120 YU2 binding to CD4 (Fig. 8a) and 17b monoclonal antibody (MoAb) (Fig. 8b) [13]. Increasing concentrations of 12p1 inhibited the binding of YU2 to both CD4 and 17b surfaces. The fraction of the initial rate of YU2 binding in the presence or absence of peptide was determined for each peptide concentration and plotted as a function of the log of the peptide concentration (Fig. 8c).

8 Epitope Mapping

Epitopes are specific sites on antigen molecules that are recognized by antibodies. Structural studies showed that around 15 amino acid residues are often associated with each epitope [14]. Epitope mapping on the biosensor can be used to characterize both antigens and MoAbs. Compared to the labeling requirements for EIA or RIA, using biosensor technology brings many advantages such as fast response, simple performance of experiments, and low sample consumption. Typical epitope mapping experiments on the biosensor involve immobilizing the primary antibody on the surface, then capturing the antigen and testing whether a secondary antibody is capable of binding to the antigen. Using this method it is possible to screen a variety of MoAbs specific to different epitopes presented on the antigen. In the following example (Fig. 9), different MoAbs against PR3 serine protease were tested [15]. First, antibody PR3G-2 was captured on a rabbit anti-mouse IgG1. The rest of the binding sites were blocked with IgG MoAb. PR3 was captured on PR3G-2 and antibodies specific to different epitopes of PR3 were tested for binding. A total of 13 MoAbs were studied for binding to PR3. To examine whether MoAbs to PR3 compete in their binding to PR3, all MoAbs were tested pair-wise. MoAbs inhibiting binding of secondary MoAbs resulting in binding $< 10\%$ of maximal binding were assumed to be MoAbs that recognize similar or closely related epitopes. MoAbs decreasing binding of secondary

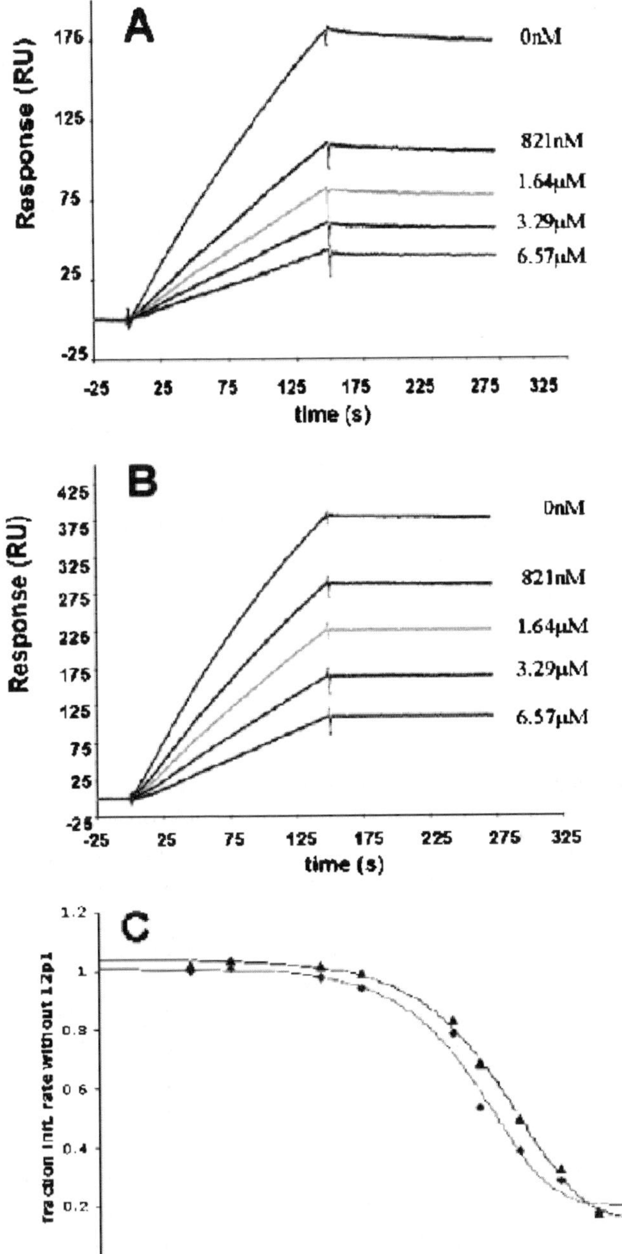
A
Response (RU)
176
125
75
25
-25
0nM
821nM
1.64μM
3.29μM
6.57μM
-25 25 75 125 175 225 275 325
time (s)
B
Response (RU)
425
375
325
275
225
175
125
75
25
-25
0nM
821nM
1.64μM
3.29μM
6.57μM
-25 25 75 125 175 225 275 325
time (s)
C
fraction init. rate without 12p1
1.2
1
0.8
0.6
0.4
0.2
0
-9 -8 -7 -6 -5
log [12p1] (M)

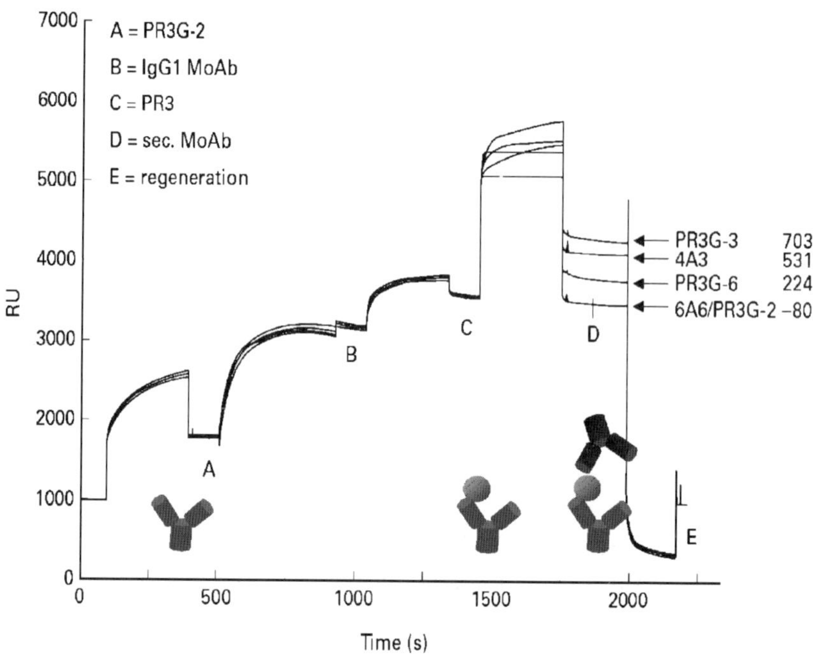

Fig. 9 Overlay plot of sensorgrams showing the binding of the MoAbs PR3G-3, 4A3, PR3G-6, 6A6, and PR3G-2 on PR3 presented by PR3G-2. *A* 800 RU binding of PR3G-2 on a rabbit anti-mouse IgG1 (RAM-IgG1) containing sensor chip. *B* Unoccupied RAM-IgG1 sites were occupied by a control IgG1 MoAb (1400 RU). *C* 400 RU of captured PR3. *D* Binding of PR2G-3, 4A3, PR3G-6, 6A6, and PR3G-2 to PR3. RU values shown represent the increase in RU from point *C* to *D*. After regeneration RU came back to base level (*E*). Reproduced from [15] with permission from Blackwell Publishing © 1999

MoAbs from 10 to 50% of maximal binding were assumed to be MoAbs recognizing overlapping epitopes. Pairs of MoAbs decreasing binding less than 50% of maximal binding were assumed to recognize different epitopes of PR3.

9 Binding Stoichiometry

Because SPR responses are generally proportional to mass, it is possible to gain some insight into the stoichiometry of an interaction by comparing the amount of ligand immobilized to the amount of analyte it can bind. If one knows the mass ratio of the two species then a stoichiometry of interaction can be calculated according to the following equation:

$$\text{stoichiometry} = \frac{R_{\text{max}} \times M_{\text{wL}}}{M_{\text{wA}} \times R_{\text{L}}},$$

where R_{max} is maximum capacity for analyte binding, and R_L is the density of immobilized ligand. Binding stoichiometry is often used as a rough indicator of the integrity of immobilized ligand. Usually, if the stoichiometry is significantly less than expected it is indication of the loss of ligand activity. It is, however, not possible to determine if the loss has come from the immobilization process itself or if the starting material was not fully active. Therefore, stoichiometric measurements based on capturing methods rather than on direct immobilization are often more reliable. As an example, we captured a dimeric Fc-fusion construct of the erythropoietin receptor using a protein A surface and, based on the RU levels, could demonstrate that it bound only one ligand [16].

10 Lipid Surfaces

SPR experiments are not limited to studying soluble protein systems. One of the most significant advancements recently has been the development of methods to create lipid surfaces on the biosensor. One application involves the study of how drugs interact with lipid surfaces as a measure of membrane transport properties. SPR-based biosensors represent a powerful tool for studying this kind of interaction. The L1 chip available for Biacore instruments is specially designed with highly hydrophobic groups on a dextran layer to facilitate capture of liposomes and micelles. Liposomes are hydrated particles of lipids that can be prepared by extrusion [17]. Liposomes have been shown to remain as intact spheres on the sensor surface while micelles prepared by mixing lipids with detergent have been shown to form confluent bilayers (Fig. 10). This allows investigators to choose the type of lipid environment for their studies.

An example of a sample analysis cycle using the L1 chip for liposome capture is shown in Fig. 11a. These authors used a Biacore S51 instrument to

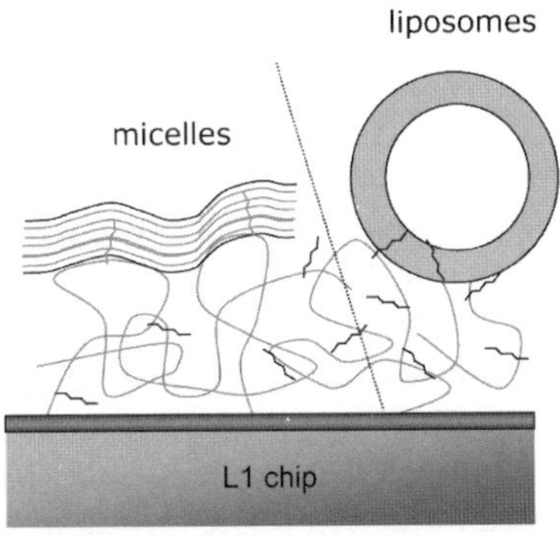

Fig. 10 Scheme of L1 chip with captured micelles and liposomes

Fig. 11 **a** Sensorgram showing the injection sequence for analysis of one compound. ▶ An analysis cycle consisted of: 3-min injections of each liposome; wash of the flow system (except the sensor surface) with the regeneration solution; a 1-min injection of running buffer to check for carryover effects from the previous cycle; a 1-min injection of compound over the reference and liposome spots; and finally, two injections of regeneration solution to wash off liposomes from the surface. A flow rate of 30 μL min^{-1} was used throughout the assay except for the liposome injections, which were performed at 10 μL min^{-1}. **b** Relative responses on Avanti-blend liposomes for drugs from four experiments. Response values were normalized against the negative control ceftriaxone, set at response 0, standard deviation (stdev) 0.2 and propranolol, set at response 100, stdev 2.1. $n = 20$ for the controls (*striped bars*), $n = 8$ for lactulose and raffinose, and $n = 4$ for all other compounds. Reproduced from [18] with permission from John Wiley and Sons © 2005

study binding of 78 compounds to liposomes POPC and Avanti-blend captured on sensor chip [18]. Using an automatized SPR-based biosensor it is possible to develop a screening method to rank and characterize the binding of compounds to captured liposomes of different properties based on binding responses, as shown in the histogram (Fig. 11b).

Micelles have been used to build lipid bilayers around membrane proteins captured on sensor surface to mimic their natural environment. A good example of this type of application involves the study of G-protein coupled receptors (GPRCs). Receptor was captured on an L1 chip surface using an immobilized antibody that recognizes an additional tag presented on the receptor followed by injection of micelles to form the lipid bilayer [19] (Fig. 12). Others have used this approach to monitor interactions of ligands and G-proteins with GPRCs [20]. This approach may provide a general method for studying a variety of membrane-associated systems and even ion channels.

11 Screening Methods

Automatization is one of the most powerful features of advanced biosensor technology. Using Biacore instruments it is possible to automatically run hundreds of assays per day. This feature also provides a big advantage when studying membrane-associated receptors (Fig. 13). Using automated instruments it is possible to automatically screen a variety of solubilization conditions for membrane-associated receptors and determine conditions that can maintain the receptor in the most active conformation for suitable conformation-sensitive probes (conformation-dependent MoAbs, chemokines) [21]. In the following example, CCR5 receptor was captured on antibody surface and tested for activity. The throughput of this method allows screening of at least 50 conditions per day. Using the autosampler to perform

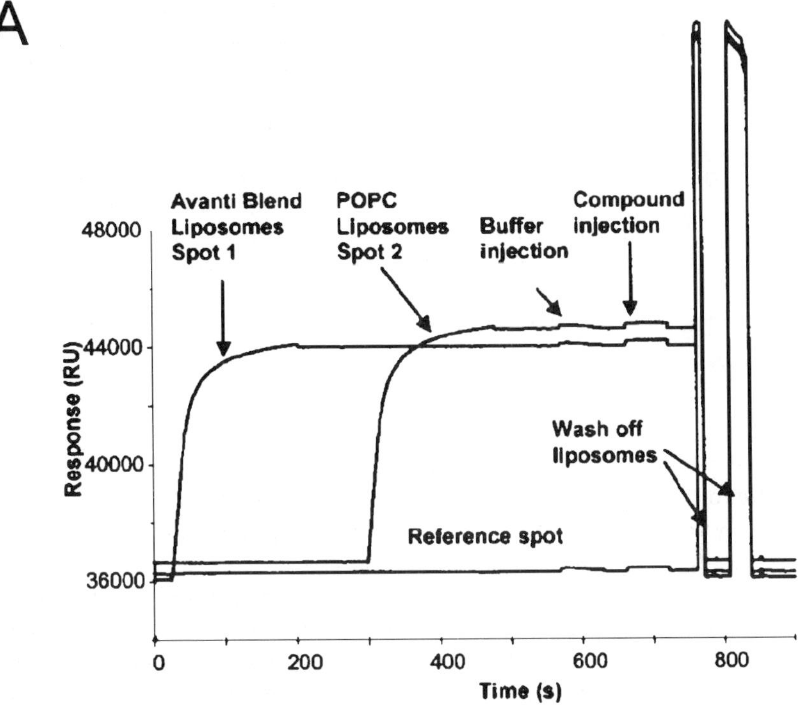
A
Avanti Blend Liposomes Spot 1
POPC Liposomes Spot 2
Buffer injection
Compound injection
Wash off liposomes
Reference spot
Response (RU)
48000
44000
40000
36000
0
200
400
600
800
Time (s)

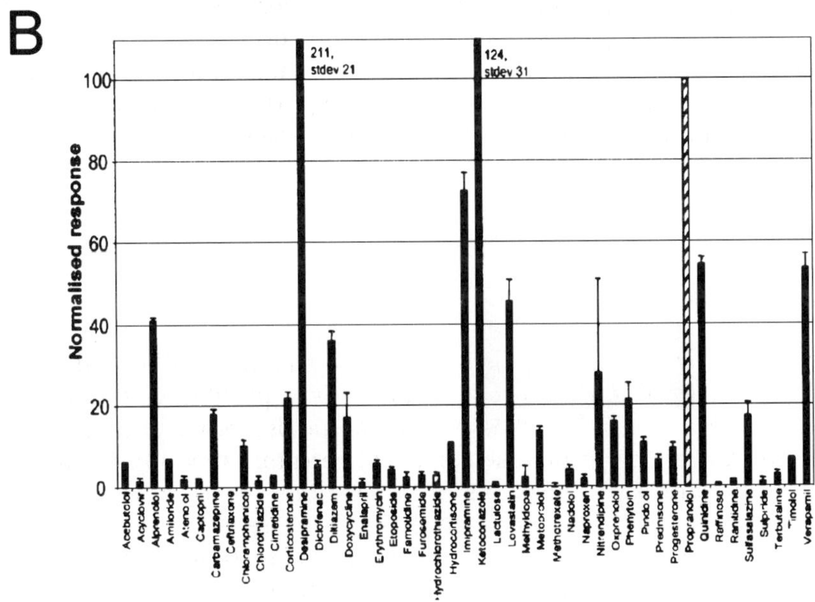
B
Normalised response
211, stdev 21
124, stdev 31
100
80
60
40
20
0
Acebutolol
Acyclovir
Alprenolol
Amiloride
Atenolol
Captopril
Carbamazepine
Ceftriaxone
Chloramphenicol
Chlorothiazide
Cimetidine
Corticosterone
Desipramine
Diclofenac
Diltiazem
Doxycycline
Enalapril
Erythromycin
Etoposide
Famotidine
Furosemide
Hydrochlorothiazide
Hydrocortisone
Imipramine
Ketoconazole
Lactulose
Lovastatin
Methyldopa
Metoprolol
Methotrexate
Nadolol
Naproxen
Nitrendipine
Oxprenolol
Phenytoin
Pindolol
Prednisone
Progesterone
Propranolol
Quinidine
Raffinose
Ranitidine
Sulfasalazine
Sulpiride
Terbutaline
Timolol
Verapamil

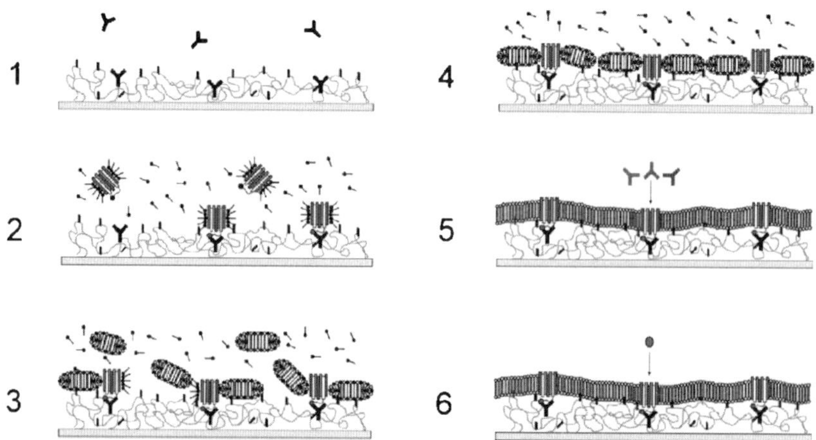

Fig. 12 Scheme of capture and reconstitution method. *1* Immobilize a capturing molecule that recognizes the GPCR distant from the ligand-binding site. In contrast to immobilizing on a standard dextran chip (CM5), the immobilization is done using an L1 chip consisting of a dextran surface that contains hydrophobic alkane groups. *2* Capture a detergent-solublized GPCR to the immobilized biomolecule on the surface. *3* Reconstitute a lipid bilayer around the receptor by injecting lipid/detergent mixed micelles across the surface. *4* Wash the surface with buffer to dissociate the detergent from the micelles, leaving behind a lipid bilayer. *5* Test the functional properties of the membrane-associated GPCR using conformation-dependent antibodies. *6* Study the binding of various ligands to the captured GPCRs. Reproduced from [19] with permission from Elsevier © 2003

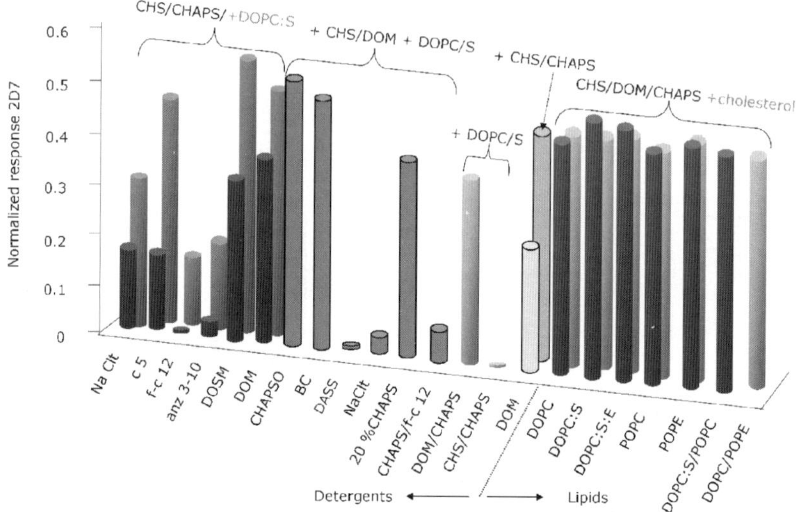

Fig. 13 Screening of CCR5 solubilization using different detergent/lipid combinations. Normalized binding responses for conformation-dependent antibody (2D7, 156 nM). Reproduced from [21] with permission from Elsevier © 2005

the mixing steps of cells with the solubilization buffer and the detergent/lipid mixture, it is possible to normalize for solubilization time. Compared to standard laborious solubilization methods that have to be performed in a cold room with a limited number of samples, automatized solubilization brings many advantages: a large number of samples can be measured, easy set up of experiment, and standardized control of the whole solubilization procedure.

12
Small Molecules

The sensitivity of Biacore instruments allows them to be used to detect low-molecular-weight molecules such as 200 Da binding to surface immobilized macromolecules (> 100 kDa). Some small compounds are not generally soluble in common buffers and addition of organic solvents like DMSO is necessary to achieve their solubility. However, these organic solvents are characterized by a high refractive index that can influence binding data for compounds caused by the excluded volume effect. Due to the volume occupied by an immobilized protein, the reaction flow-cell will have a lower baseline response than the reference flow-cell (lacking immobilized protein) in the presence of DMSO. When the data generated across the reference flow-cell is subtracted from that generated across the reaction flow-cell, the result will be negative, which may mask the actual binding of small molecules. To correct for this effect, a series of samples containing different DMSO concentrations should be injected from which a standard curve is constructed. The

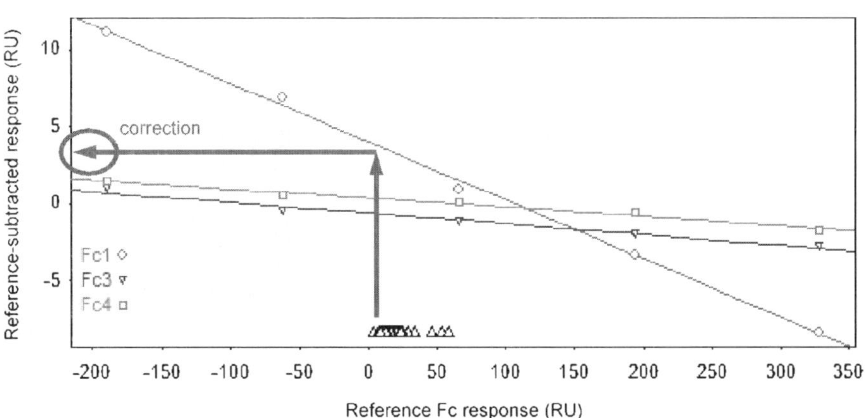

Fig. 14 Example of DMSO calibration curve analyzed using Scrubber software. Responses for standards obtained for reaction flow-cells are plotted versus responses obtained for reference flow-cell. Linear fit is used to fit the data. The *black open triangles* along the x-axis indicate where the reference Fc values lie for each analyte injection cycle

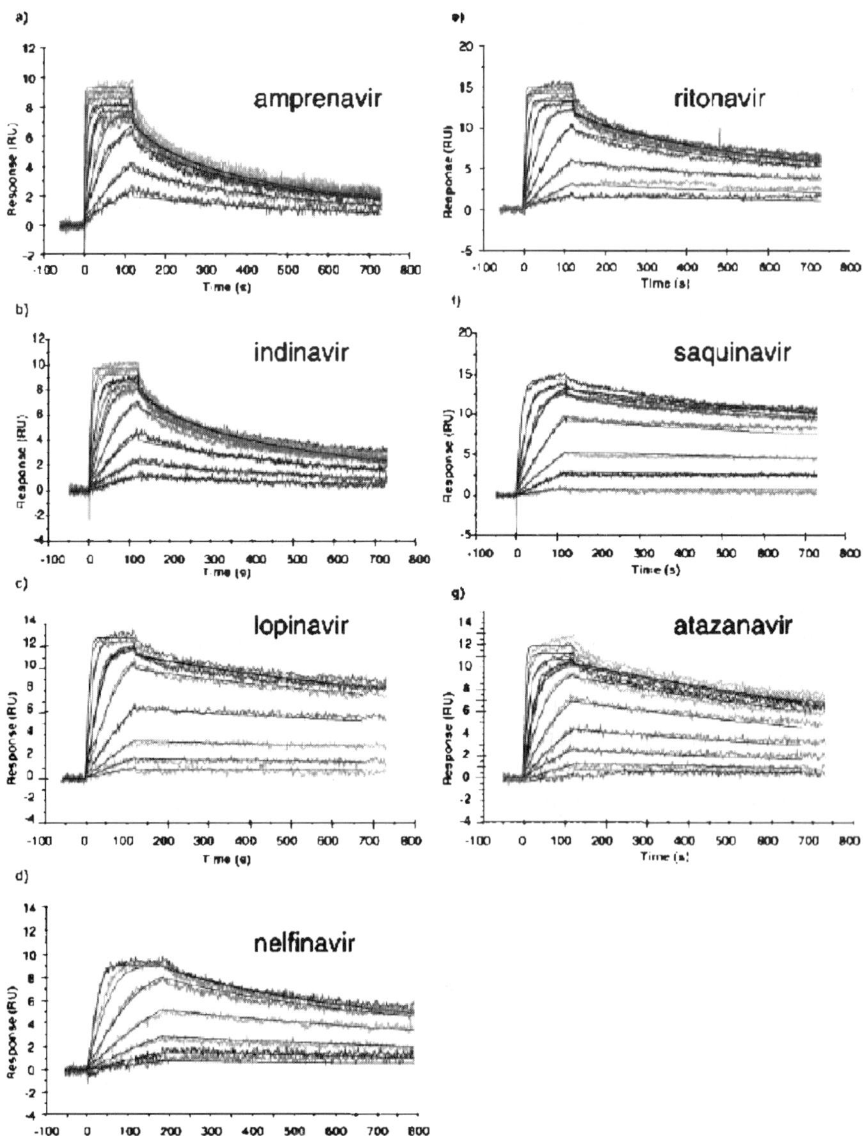

Fig. 15 Experimental (*grayscale traces*) and modeled sensorgrams (*solid black lines*) for the interaction between HIV_1 protease inhibitors (in twofold serial dilutions) at 25 °C. The modeled sensorgrams were based on the kinetic parameters obtained from global analysis of the experimental data using a 1 : 1 binding model accounting for mass transport. **a** Amprenavir (1.6–400 nM); **b** Indinavir (1.6–400 nM); **c** Lopinavir (0.4–50 nM); **d** Nelfinavir (3.2–200 nM); **e** Ritonavir (1.6–400 nM); **f** Saquinavir (1.6–200 nM); and **g** Atazanavir (0.4–200 nM). Reproduced from [11] with permission from John Wiley and Sons © 2004

DMSO concentrations of the calibration series must span the refractive index increments in the samples. A calibration curve is then constructed as a dependence of responses obtained for reaction flow-cells versus responses for reference flow-cell (Fig. 14). A linear or polynomial fit is performed to correct the measured data.

A typical example of small molecule analysis with SPR is the study of enzyme–inhibitor complexes. In the following experiment (Fig. 15), the authors measured the affinity of various HIV-protease inhibitors [11] using a Biacore S51 instrument. These high-quality data illustrate a significant new application of SPR technology that is making an impact in drug discovery.

13 Summary

SPR-based biosensors represent a real-time and label-free technology that is heavily utilized by both basic researchers and within the pharmaceutical industry. Biosensors can support an incredible range of applications from qualitative binding to high-resolution kinetic analysis. Nearly any interaction involving biological systems (including low-molecular-weight components, proteins, nucleic acids, and even lipid surface environments) are amenable to these instruments. In the future, we will see SPR-based technology continue to expand with advancements in higher-throughput and higher-sensitivity platforms. What was once an obscure and niche technology has now become an essential component of protein analysis and has been adopted as a mainstream technology.

References

1. Myszka DG (1999) J Mol Recognit 12:279
2. Sjolander S, Urbaniczky C (1991) Anal Chem 63:2338
3. Glaser RW (1993) Anal Biochem 213:152
4. Karlsson R, Roos H, Fagerstam L, Persson B (1994) Methods 6:99
5. Fisher RJ, Fivash M, Casas-Finet J, Bladen S, McNitt KL (1994) Methods 6:121
6. Myszka DG, Morton TA (1998) Trends Biochem Sci 23:149
7. Myszka DG, Morton TA, Doyle ML, Chaiken IM (1997) Biophys Chem 64:127
8. Binz HK, Amstutz P, Kohl A, Stumpp MT, Briand C, Forrer P, Grutter MG, Pluckthun A (2004) Nat Biotechnol 22:575
9. Johansson C, Finger LD, Trantirek L, Mueller TD, Kim S, Laird-Offringa IA, Feigon J (2004) J Mol Biol 337:799
10. Biskup C, Bohmer A, Pusch R, Kelbauskas L, Gorshokov A, Majoul I, Lindenau J, Benndorf K, Bohmer FD (2004) J Cell Sci 117:5165
11. Shuman CF, Hamalainen MD, Danielson UH (2004) J Mol Recognit 17:106
12. Gregory LA, Thielens NM, Matsushita M, Sorensen R, Arlaud GJ, Fontecilla-Camps JC, Gaboriaud C (2004) J Biol Chem 279:29391

13. Biorn AC, Cocklin S, Madani N, Si Z, Ivanovic T, Samanen J, Van Ryk DI, Pantophlet R, Burton DR, Freire E, Sodroski J, Chaiken IM (2004) Biochemistry 43:1928
14. Johne B (1998) Mol Biotechnol 9:65
15. Van Der Geld YM, Limburg PC, Kallenberg CGM (1999) Clin Exp Immunol 118:487
16. Morton T, Myszka DG (1998) Kinetic analysis of macromolecular interactions using surface plasmon resonance biosensors. Methods Enzymol 295:256
17. www.avantilipids.com
18. Frostell-Karlsson A, Widegren H, Green CE, Hamalainen MD, Westerlund L, Karlsson R, Fenner K, van de Waterbeemd H (2005) J Pharm Sci 94:25
19. Stenlund P, Babcock GJ, Sodroski J, Myszka DG (2003) Anal Biochem 316:243
20. Karlsson OP, Lofas S (2002) Anal Biochem 300:132
21. Navratilova I, Sodroski J, Myszka DG (2005) Anal Biochem 339:271

Springer Ser Chem Sens Biosens (2006) 4: 177–190
DOI 10.1007/5346_019

Published online: 5 July 2006

SPR Biosensors for Detection of Biological and Chemical Analytes

Jakub Dostálek[1] · Jon Ladd[2] · Shaoyi Jiang[2] · Jiří Homola[1] (✉)

[1]Institute of Radio Engineering and Electronics, Prague, Czech Republic
homola@ure.cas.cz

[2]Department of Chemical Engineering, University of Washington, Box 351750, Seattle, WA 98195-1750, USA

Keywords Biosensor · Detection format · Environmental monitoring · Fluidic unit · Food safety · Medical diagnostics · Mobile sensor · Portable sensor · Recognition element · Sample preparation · Sensor · SPR · SPR optical platform · Surface plasmon resonance

1 Introduction

Surface plasmon resonance (SPR) biosensors present a mainstay technology for research of macromolecules and their interactions in life sciences and pharmaceutical research. In addition, SPR biosensors hold potential for many other applications of paramount importance, including detection of contaminants related to environmental monitoring, human health indicators for medical diagnostics, and foodborne pathogens and toxins implicated in food safety and security. Existing commercial SPR biosensors are not designed for in-field detection or continuous monitoring of chemical and biological analytes. In order to address analytical needs in theses areas, development of SPR biosensors suitable for out-of-laboratory applications and analysis of complex real-world samples is pursued in research laboratories worldwide.

2 Concept of SPR Biosensor System for Field Use

The SPR biosensor systems for analysis of complex samples in the field have to integrate several key elements. These include, in particular, a *sample preparation unit*, a *fluidic system*, a *biorecognition element*, and an *SPR optical platform* (Fig. 1).

In this biosensor system, a sample is pretreated in the sample preparation unit and delivered by the fluidic system into contact with the biorecognition element immobilized on the sensor surface. The SPR optical platform converts its specific interaction with the analyte into the sensor output.

This chapter is devoted to a description of the state of the art in the development of these key elements and their integration for SPR biosensor instruments for field use.

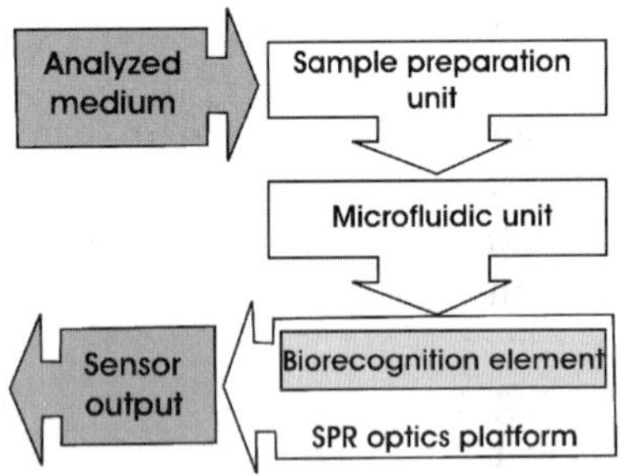

Fig. 1 Scheme of key modules supporting an SPR biosensor instrument

3 Sample Preparation Unit

SPR biosensors are devices that are suitable for analysis of aqueous samples. Therefore, in order to detect target analytes in different real-world matrices (e.g., tissue, meat, soil, and air) the analyte has to be transferred to a liquid by a sample preparation unit. Numerous sample pretreatment methods for gas, solid, and crude liquid samples compatible with SPR biosensors are available. For detection in gas environments such as air, real-time trapping of analyte into an aqueous solution is possible by using collectors such as a wetted-wall cyclone particle collector [1]. Several optical biosensors have been integrated with these collectors and installed on aerial vehicles for real-time detection

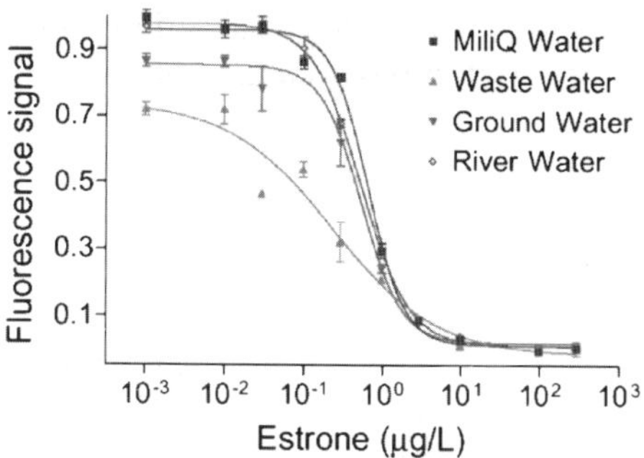

Fig. 2 Changes in the calibration of a fluorescence-based biosensor for detection of estrone due to the non-specific interaction of biorecognition elements with residual matrix components [9]

of airborne substances [2, 3]. Solid matrices such as tissue or soil are usually homogenized and suspended in a buffer or solvents. Then, the matrix separation is performed by filtration [4], centrifugation [5], or immunomagnetic separation [6]. Liquid samples, either collected directly or obtained as supernatants from solid or gaseous matrices, can be analyzed with an SPR biosensor directly.

Additional sample treatment is necessary prior to sample injection into an SPR biosensor when crude samples are analyzed (e.g., blood, waste water, or supernatants from solid matrices). In these samples, residual matrix components can non-specifically interact with the biomolecular recognition elements (e.g., dissolved organic carbon [7]) and variations in sample properties such as pH and ionic strength can alter the specific interaction between an analyte and a biomolecular recognition element. The matrix effects can lead to variations in sensor calibration [8–10] resulting in false sensor responses. In order to reduce these effects, analyzed aqueous samples can be buffered to stabilize their pH and ionic strength and filtered to remove residual matrix components. Figure 2 illustrates the effect of matrix composition. Calibration curves are shown for an estrone fluorescence-based biosensor performing detection in water samples from different sources (samples were buffered prior to their analysis).

4 Fluidic Unit

In SPR biosensors, a fluidic unit is necessary to provide precise control of sample delivery to the sensor surface as the amount of analyte captured by the biorecognition elements (and thus the sensor response) depends on the flow conditions at the sensor surface [11, 12].

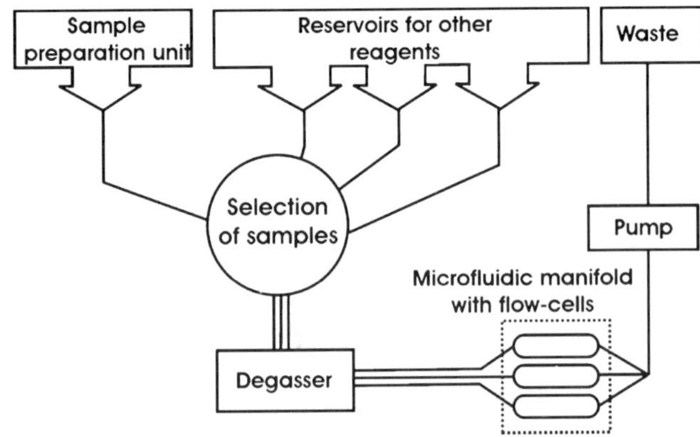

Fig. 3 Scheme of a typical fluidic unit supporting an SPR biosensor

In general, the fluidic unit needs to encompass reservoirs to contain analyzed liquids and other reagents (e.g., solutions for sensor regeneration and sensor surface washout), a pump to flow these liquids through the sensor, valves to control their injection, and flow-cells for their distribution on the sensor chip (Fig. 3). Peristaltic pumps are frequently used to flow liquid samples through the sensor flow-cell [2, 13]. Selection of samples and reagents is typically achieved with selection valves [2] which can be substituted with air vents placed within reservoirs [13]. The fluidic channels and flow-cells can be produced by combining conventional machining (e.g., drilling of input and output ports) with microfabrication of the fluidic manifold using technologies such as molding of plastics [14], casting in poly(dimethylsiloxane) [15, 16], and laser cutting of thin polymer layers [17, 18]. The fluidic system can be combined with degassers for the removal of air dissolved in a sample as it can produce air bubbles at the sensor surface interfering with the SPR biosensing [2].

To date, several SPR biosensor systems with integrated automated fluidic system have been reported [2, 19, 20]. However, these devices rely on bulky components (e.g., external pumps and valves), which limits their further miniaturization. In future, we expect that development of more compact fluidic units will benefit from current advances in the micropumps and microvalves [21] and microfluidic technologies pursued for Micro Total Analysis Systems (μTAS) and Lab-on-a-Chip devices [22–24].

5 SPR Optical Platform

In order to create a portable/mobile SPR biosensor for applications in the field, easy-to-use SPR biosensor instruments that can deliver high accuracy detection in real-world environmental conditions need to be developed. To meet these requirements, these biosensor instruments have to encompass a robust and compact SPR optical platform providing reference channels for the compensation of fluctuations in SPR sensor response due to changes in optical properties of the analyzed sample and variations of environmental conditions. Moreover, to enable simultaneous detection of multiple analytes, SPR optical platforms have to support multiple independent sensing channels.

A large variety of SPR optical platforms have been developed (see Chap. 4 in this volume [63]). The SPR sensors allowing the highest degree of miniaturization of the SPR optics are based on optical fibers [25–28]. These sensors have potential for localized detection including in vivo diagnostic applications. However, the fiber optic SPR sensors exhibit a limited accuracy, which up to now has hindered their applications for detection of chemical and biological analytes. In order to provide more accurate SPR sensor instruments, several miniaturized SPR optical platforms relying on bulk optics and the attenuated total reflection (ATR) method have been developed. These include the SPR platform based on angular modulation of SPR proposed by Elkind et al. [29], Kawazumi et al. [30] and Thirstrup et al. [31]. Another compact SPR sensor platform based on the wavelength modulation of SPR and wavelength division multiplexing of sensing channels [17] has been recently developed (Fig. 4).

SPR sensor platforms supporting easy-to-interchange sensor chips are desired to allow fast and simple replacement of chips or introduction of a sensor chip with a desired biomolecular recognition element. In a majority of the

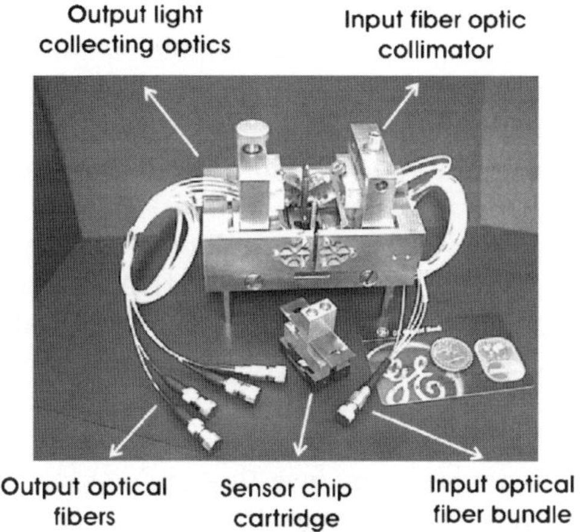

Fig. 4 Compact optical bench supporting an eight-channel SPR sensor relying on wavelength modulation of SPR and WDM multiplexing of sensing channels; developed at the Institute of Radio Engineering and Electronics, Prague

current SPR biosensor platforms an optical contact of the sensor chip with a coupling prism needs to be established. The optical contact is typically achieved by means of a refractive index matching oil or a soft polymer which, for in-field applications, makes the loading of a sensor chip rather inconvenient. Therefore, SPR optical platforms based on light-pipes [31, 32] and diffraction gratings [33] have been proposed to avoid the necessity of establishing an optical contact.

SPR biosensors for analyzing complex matrices in realistic environmental conditions need to discriminate between the refractive index changes due to specific interaction with an analyte and those due to background refractive index variations. Fluctuations in the background refractive index are typically caused by changes in composition of the sample (e.g., residual matrix components) and by temperature variations. The discrimination between these changes can be achieved by using reference channels [32] or by means of decomposition of SPR variations using multiwavelength spectroscopy of surface plasmons [17, 34–36]. The effect of temperature changes, which can affect the interaction of the biorecognition element with analyte as well as the performance of optical components (e.g., lightsource spectrum and detector efficiency are a function of temperature), can be reduced by stabilizing the temperature of the SPR optical platform [20].

6
Molecular Recognition Element

Numerous biorecognition elements and methods for their attachment to surfaces can be used with SPR biosensors (see Chap. 5 in this volume [64]). In applications of SPR biosensors for detection of chemical and biological analytes, biorecognition elements and their immobilization have to be selected with respect to desired specificity (detection of individual molecules or biological activity of overall sample), mode of operation (continuous monitoring or rapid detection), stability and storability (long term storage and operation in realistic environmental conditions).

SPR biosensors relying on a variety of biorecognition elements (including antibodies [37–40], hormone receptors [41, 42], and whole cells [43]) have been used for detection of analytes. Among these, antibody biorecognition elements are the most popular due to their high affinity, versatility, and commercial availability [44]. As an alternative to biorecognition elements, other receptors such as molecular imprinted polymers (MIPS) [45] and organic synthetic receptors [46] were investigated due to their potential higher stability in environmental conditions. However, to date the accuracy and specificity these SPR sensors are still significantly lower then their biorecognition element-based counterparts. In continuous monitoring SPR biosensors, biorecognition elements with lower affinity are preferred for achieving reversible interaction with target analyte [47]. SPR biosensors for rapid detection employ high affinity biorecognition elements to achieve the lowest detection limits.

Immobilization of biorecognition elements presents an important challenge to sensor development. Sensor sensitivity and specificity are two concerns that influence biorecognition element immobilization techniques. Sensitivity is related to the amount of biorecognition element that is available on the surface for analyte binding. When antibodies are used as the biorecognition element, the orientation and conformation of the antibodies can vary the amount of available analyte-binding sites on the surface. Recent studies showed that orientating the analyte-binding pockets away from the surface can increase the sensitivity of SPR biosensors [48]. Specificity is another important concern for surfaces of SPR biosensors. Integration of non-fouling materials as a background for sensor surfaces has become an important focus of SPR biosensor development. Oligo (ethylene glycol) (OEG) is one material used to resist non-specific adsorptions of biomolecules. Mixed self-assembled monolayers (SAMs) consisting of a non-fouling background and a binding element have come to the forefront as a simple, yet effective means for addressing the issue of non-fouling sensor surfaces [49–52]. While these surfaces make great strides in limiting the amount of non-specific adsorption of proteins [53, 54], other surfaces resistant to bacteria and other extremely complex matrices are still needed. Recently zwitterionic SAMs were reported to show good non-fouling characteristics for both proteins and live bacteria [55].

The stability of the biorecognition elements immobilized on the sensor surface is an important factor for use of SPR biosensors in field applications. In general, long-term storage and exposure to environmental conditions could decrease the functionality of the immobilized biorecognition element. In SPR biosensors, protein biorecognition elements are mostly used. Storage characteristics of protein arrays were studied by, e.g., Angenendt et al. [56]. He showed that his protein arrays, consisting of five different antibodies, could be stored for a period of 8 weeks. Proteins are known to be more stable in a solution than anchored to a surface. Therefore, a novel method has been introduced to reduce the amount of time that protein biorecognition elements must stay immobilized on the sensor surface. This method implements site-directed immobilization of a protein–DNA conjugate to a surface that is modified with a single strand DNA (ssDNA) [49, 51, 57]. In this work, the protein conjugate consists of an antibody chemically linked to an ssDNA target that has a sequence complimentary to the one bound at the surface. The antibody–DNA conjugate is immobilized on the surface via sequence-specific hybridization. Using this methodology, DNA arrays that are more stable than protein arrays can be prepared and stored. Antibodies can be immobilized on such a sensor chip prior to detection from a solution. In addition to maintaining the stability of immobilized proteins, this approach offers other advantages. Sensors relying on DNA-directed immobilization of biorecognition elements have shown as much as a 50-fold increase in sensitivity over conventional protein sensors [49, 51]. Furthermore, dehybridization has been shown to be an easy and effective means to recycle the DNA sensor surface.

7 Detection Format

In SPR biosensors for detection of chemical and biological analytes, detection formats need to be chosen depending on the size of target analyte and whether detection or continuous monitoring is needed.

Detection of analytes can be performed using either direct detection methods or indirect detection methods. In the case of direct detection methods, an analyte or parts of an analyte are bound to the sensing surface producing the sensor response. Direct detection methods include direct detection of the analyte, sandwich assays, and competitive assays. In indirect detection methods, the analyte induces a change in the state of a secondary system component, which subsequently induces a sensor response. The most commonly used indirect detection method is the inhibition assay.

In a sandwich assay, as seen in Fig. 5a, one antibody is immobilized on the sensor surface. Analyte is then flowed over the sensor surface and captured by the immobilized antibody. Following analyte capture, binding of a second antibody (normally a polyclonal antibody) to the analyte at the sensor surface is measured. This amplification has a twofold effect: improvement of lower detection limits and verification of the bound analyte.

Competitive assays, as seen in Fig. 5b, are based on two analytes competing for the same recognition site at the sensor surface. One of the analytes is free and the other is typically conjugated to a larger protein, usually bovine serum albumin or casein. The concentration of the conjugated analyte is fixed from solution to solution. The two analytes are mixed in a solution and passed across the sensing surface. The sensor response will be inversely proportional to the concentration of analyte in the target solution.

In an inhibition assay, as seen in Fig. 5c, the analyzed sample is preincubated with an antibody for the targeted analyte. Subsequently, the mixture is injected in the SPR sensor with an analyte derivative immobilized on the sensor surface and the binding of the unreacted antibody to the analyte derivative is measured. As with the competitive assay, the sensor response is inversely proportional to the concentration of target analyte in the incubation solution.

Detection of medium-sized and large analytes (> 10 000 Da) is usually performed directly [37, 58]. As direct binding of low molecular weight analytes at the sensor surface does not usually produce sufficient refractive index change, they are typically detected using a competition assay [39], sandwich assay [40], or inhibition assay [38].

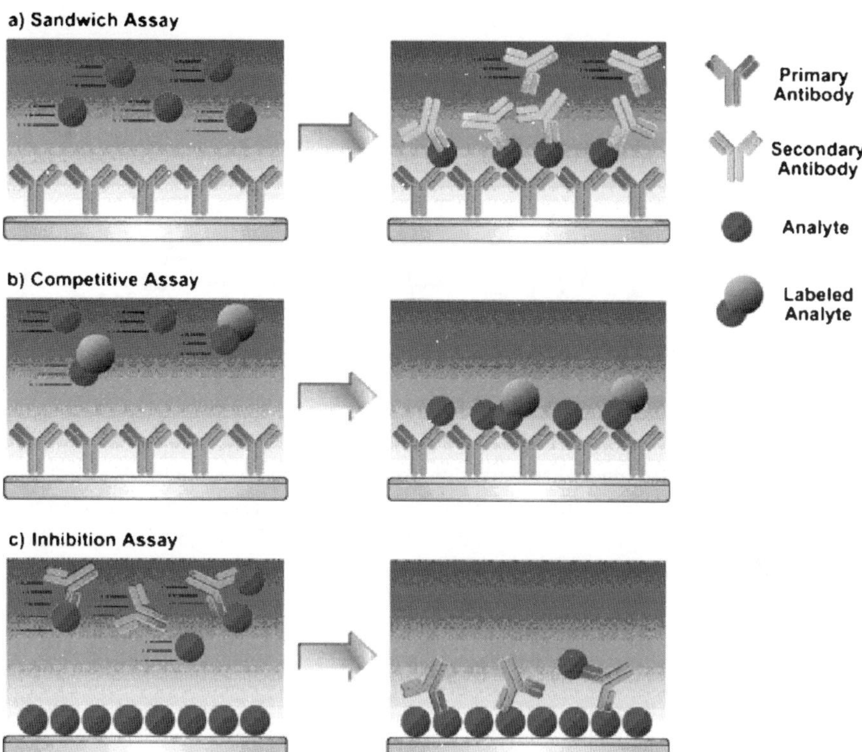

Fig. 5 Cartoon representations of three different assays typically used in detecting analytes with an SPR biosensor. **a** Sandwich assay involves the capturing of analyte by a sensing element immobilized on the sensor surface. This is followed by the binding of a secondary antibody for amplification. **b** In a competitive assay, native analyte and analyte conjugated to a larger protein compete to bind to an immobilized sensing element on the surface. **c** In an inhibition assay, analyte is incubated with a fixed concentration of antibody. This incubation solution is then passed across a surface of immobilized analyte. Free antibody binds to the sensor surface, creating an inverse relationship between concentration of analyte in the sample and sensor response

SPR biosensors for rapid detection of chemical and biological analytes usually use direct or indirect assays in conjunction with high-affinity biorecognition elements. For these elements, their interaction with an analyte is, under normal conditions, irreversible. Regeneration of the sensor surface for its repeated use can be performed by changing pH [59], using detergents [45], or with enzymes [38, 60] by which analyte bound to the biorecognition element is released leaving the sensor available for subsequent measurements (Fig. 6). In SPR immunosensors, typically tens of regeneration–detection cycles are possible without significant reduction of activity of the biorecognition elements [59, 61, 62].

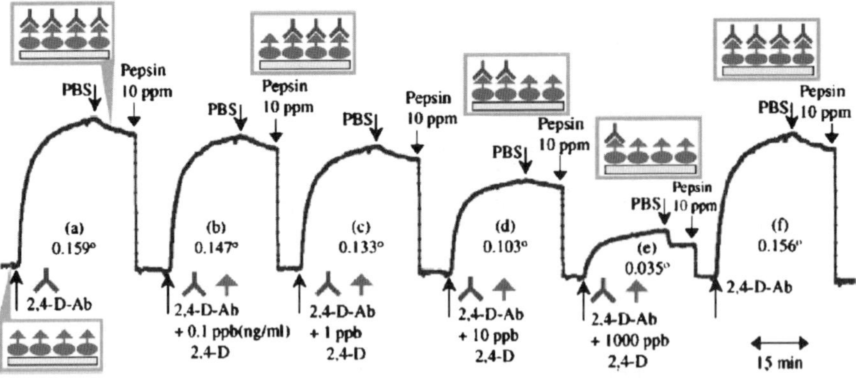

Fig. 6 SPR biosensor for detection of 2-4 dichlorophenoxyacetic acid (2,4-D) relying on inhibition assay and antibodies irreversibly interacting with the analyte; sensorgram obtained for several detection and regeneration cycles [62]

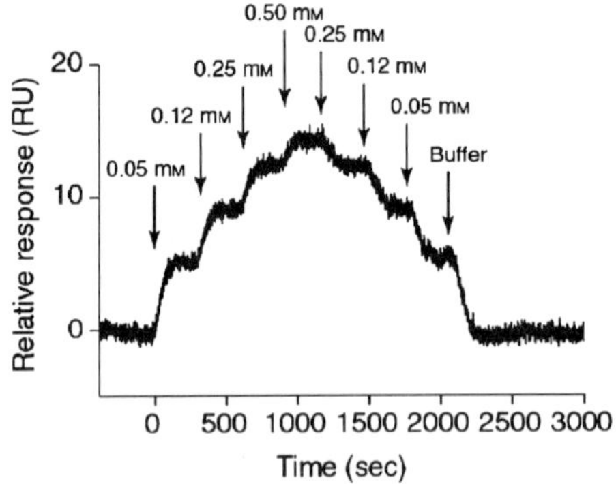

Fig. 7 SPR biosensor for continuous monitoring of maltose using direct detection of analyte and weak-affinity antibodies [47]

Additionally, SPR biosensor technology can be used for continuous monitoring of analytes. This performance can be achieved by using biorecognition elements interacting reversibly with target analyte. This type of sensor was investigated by Ohlson [47], who demonstrated continuous monitoring of maltose using weak-affinity antibodies and direct detection of analyte (Fig. 7).

8
Integration of SPR Biosensor System

Over the last few years, we have witnessed development of several portable SPR sensor instruments aimed for field applications. Based on SPREETA SPR

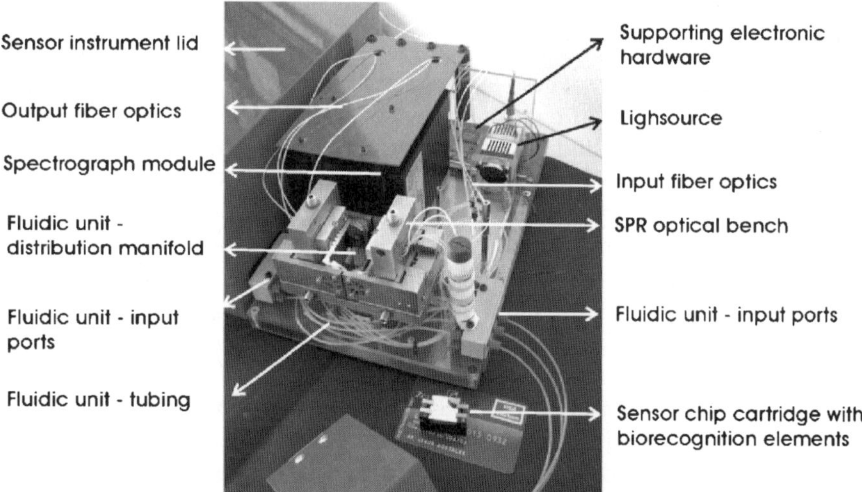

Fig. 8 Portable SPR sensor system developed at the Institute of Radio Engineering and Electronics, Prague with SPR optical platform, fluidic unit, temperature stabilization, and supporting electronic hardware

sensor (commercialized by Texas Instruments) two prototype portable sensors with one and two channels were reported by Sessay et al. [19] and Naimushin et al. [20], respectively. These portable devices encompassed SPR optics, electronic hardware, and basic fluidic systems. Currently, based on this platform Naimushin et al. developed a sensor system for detection of airborne analytes [2, 20]. This device was equipped with a sample preparation unit for collecting of analyte from aerosols. Using an aerial vehicle it was applied for measuring the spatial distribution of a model analyte (ovalbumin and horseradish peroxidase) dispersed in the atmosphere using a sandwich assay. An eight-channel portable SPR biosensor system has been recently developed at the Institute of Radio Engineering and Electronics, Prague based on a compact SPR optics bench, depicted in Fig. 4. This system (Fig. 8) incorporates a temperature-stabilized SPR optical platform, a fluidic unit, and supporting electronics. The SPR optical platform takes advantage of a special sensor chip cartridge that does not require optical matching and can be easily plugged into the sensor.

9 Summary and Outlook

In the last decade, we have witnessed a concerted research and development effort to bring SPR biosensor technology to the field and meet the need for the rapid detection and identification of chemical and biological substances in important areas such as medical diagnostics, environmental monitoring, food safety, and security. These applications present SPR sensor technology with unique challenges in terms of complexity and diversity of sample matrices (gaseous, liquid, and solid samples), type of deployment (mobile or portable versus permanently installed sensor system), detection environment (field, mobile laboratory, industrial plant, etc.), and mode of operation (rapid detection versus continuous monitoring). To address these challenges, the SPR sensor systems have to integrate multiple key functions such as sample collection and preparation, sample delivery, capture of analyte from the sample by biomolecular recognition elements, and measurement of the amount of captured analyte using the SPR method.

In recent years, first prototypes of SPR biosensor systems integrating these elements have been reported and their application for detection of biological analytes in the field have been demonstrated. Undoubtedly, advances in the development of the key elements – sample preparation technology, microfluidics, biomolecular recognition elements, SPR optical platform – will further stimulate this effort and eventually lead to in-field SPR sensor systems becoming a commercial reality.

Acknowledgements This work was supported by grants from the United States Food and Drug Administration (contract FD-U-002250) and the European Commission (contract QLK4-CT-2002-02323).

References

1. Bergman W, Shinn J, Lochner R, Sawyer S, Minalovich F, Mariella R (2005) J Aerosol Sci 36:619
2. Naimushin A, Spinelli C, Soelberg S, Mann T, Stevens R, Chinowsky T, Kauffman P, Yee S, Furlong C (2005) Sensor Actuator B Chem 104:237
3. Anderson GP, King KD, Cuttino DS, Whelan JP, Ligler FS, MacKrell F, Bovais CS, Indyke DK, Foch RJ (1999) Field Anal Chem Technol 3:307
4. Homola J, Dostálek J, Chen S, Rasooly A, Jiang S, Yee SS (2002) J Microbiol 75:61
5. Strong A, Stimpson DI, Bartholomew DU, Jenkins TF, Elkind JL (1999) SPIE 3710:362
6. Shiver-Lake LC, Shubin YS, Ligler FS (2003) J Food Protect 66:1851
7. Hennion M-C, D. B (1998) Anal Chim Acta 362:3
8. Rowe-Tait CA, Hazzard JW, Hoffman KE, Cras JJ, Golden JP, Ligler FS (2000) Biosens Bioelectron 15:579
9. Rodriguez-Mozaz S, Reder S, Lopez de Alda M, Gauglitz G, Barceló D (2004) Biosens Bioelectron 19:633
10. Tschmelak J, Proll G, Gauglitz G (2004) Biosens Bioelectron 20:743
11. Myszka DG, He X, Dembo M, Morton TA, Goldstein B (1998) Biophys J 78:583
12. Myszka DG, Morton TA, Doyle ML, Chaiken IM (1997) Bioanal Chem 64:127
13. Dodson JM, Feldstein MJ, Leatzow DM, Flack LK, Golden JP, Ligler FS (2001) Anal Chem 73:3776

14. Sjölander S, Urbanitzky C (1991) Anal Chem 63:2338
15. Goodrich TT, Lee HJ, Corn RM (2004) J Am Chem Soc 126
16. Wheeler AR, Chah SC, Whelan RJ, Zare RN (2004) Sensor Actuator B Chem 98:208
17. Dostálek J, Vaisocherová H, Homola J (2005) Sensor Actuator B Chem 108:758
18. Dostálek J, Homola J, Miler M (2005) Sensor Actuator B Chem 107:154
19. Sesay A, Cullen D (2001) Environ Monit Assess 70:83
20. Naimushin A, Soelberg S, Bartholomew D, Elkind J, Furlong C (2003) Sensor Actuator B Chem 96:253
21. Woias P (2005) Sensor Actuator B Chem 105:28
22. Weigl BH, Bardell R, Shulte T, Battrell F, Hayenga J (2001) Biomed Microdev 3:267
23. Becker H, Gärtner C (2000) Electrophoresis 21:12
24. Verpoorte E, Rooij NF (2003) Proceedings IEEE 91:930
25. Jorgenson RC, Yee SS (1993) Sensor Actuator B Chem 12:213
26. Ronot-Trioli C, Trouillet A, Veillas C, Gagnaire H (1998) Sensor Actuator B Chem 54:589
27. Obando LA, Genteman DJ, Holloway JR, Booksh KS (2004) Sensor Actuator B Chem 100:449
28. Slavík R, Homola J, Čtyroký J (1998) Sensor Actuator B Chem 51:311
29. Elkind JL, Stimpson DI, A.A. S, D.U. B, Melendez JL (1999) Sensor Actuator B Chem 54:182
30. Kawazumi H, Gobi K, Ogino K, Maeda H, Miura N (2005) Sensor Actuator B Chem 108:791
31. Thirstrup C, Zong W, Borre M, Neff H, Pedersen H, Holzhueter G (2004) Sensor Actuator B Chem 100:298
32. Nenninger GC, Clendenning JB, Furlong CE, Yee SS (1998) Sensor Actuator B Chem 51:38
33. Lawrence CR, Geddes NJ, Furlong DN, J.R. S (1996) Biosens Bioelectron 11:389
34. Adam P, Dostálek J, Homola J (2006) Sensor Actuator B Chem 113:774
35. Homola J, Lu H, Yee SS (1999) Electron Lett 35:311
36. Homola J, Lu H, Nenninger GC, Dostálek J, Yee SS (2001) Sensor Actuator B Chem 76:403
37. Oh B, Kim Y, Lee W, Bae Y, Lee W, Choi J (2003) Biosens Bioelectron 18:605
38. Mouvet C, Harris R, Maciag C, Luff B, Wilkinson J, Piehler J, Brecht A, Gauglitz G, Abuknesha R, Ismail G (1997) Anal Chim Acta 338:109
39. Shimomura M, Nomura Y, Zhang W, Sakino M, Lee K, Ikebukuro K, Karube I (2001) Anal Chim Acta 434:223
40. Minunni M, Mascini M (1993) Anal Lett 26:1441
41. Asano K, Ono A, Hashimoto S, Inoue T, Kanno J (2004) Anal Sci 20:611
42. Usami M, Mitsunaga K, Ohno Y (2002) J Steriod Biochem Mol Biol 81:47
43. Chio JW, Park KW, Lee DB, Lee WC, Lee WH (2005) Biosens Bioelectron 20:2300
44. Mullett WM, Lai EPC, Yeung JM (2000) Methods 22:77
45. Lotierzo M, Henry O, Piletsky S, Tothill I, Cullen D, Kania M, Hock B, Turner A (2004) Biosens Bioelectron 20:145
46. Wright J, Oliver J, Nolte R, Holder S, Sommerdijk N, Nikitin P (1998) Sensor Actuator B Chem 51:305
47. Ohlson S, Jungar C, Strandh M, Mandenius CF (2000) Tibtech 18:49
48. Chen SF, Liu LY, Zhou J, Jiang SY (2003) Langmuir 19:2859
49. Boozer C, Ladd J, Chen SF, Yu Q, Homola J, Jiang SY (2004) Anal Chem 76:6967
50. Boozer C, Yu QM, Chen SF, Lee CY, Homola J, Yee SS, Jiang SY (2003) Sensor Actuator B Chem 90:22

51. Ladd J, Boozer C, Yu QM, Chen SF, Homola J, Jiang S (2004) Langmuir 20:8090
52. Nelson KE, Gamble L, Jung LS, Boeckl MS, Naeemi E, Golledge SL, Sasaki T, Castner DG, Campbell CT, Stayton PS (2001) Langmuir 17:2807
53. Ostuni E, Yan L, Whitesides GM (1999) Colloid Surf B Biointerfaces 15:3
54. Prime KL, Whitesides GM (1993) J Am Chem Soc 115:10714
55. Chen SF, Zheng J, Li LY, Jiang SY (2005) J Am Chem Soc 127:14473
56. Angenendt P, Glokler J, Murphy D, Lehrach H, Cahill DJ (2002) Anal Biochem 309:253
57. Boozer C, Ladd J, Chen SF, Jiang S (2006) Anal Chem 75:1515
58. Homola J, Dostálek J, Chen SF, Rasooly A, Jiang SY, Yee SS (2002) Int J Food Microbiol 75:61
59. Yu Q, Chen S, Taylor A, Homola J, Hock B, Jiang S (2005) Sensor Actuator B Chem 107:193
60. Gobi K, Miura N (2004) Sensor Actuator B Chem 103:265
61. Shankaran DR, Matsumoto K, Toko K, Miura N (2006) Sensor Actuator B Chem 114:71
62. Gobi KV, Tanaka H, Shoyama Y, Miura N (2005) Sensor Actuator B Chem 111–112:562
63. Piliarik M, Homola J (2006) SPR sensor instrumentation. In: Homola J (ed) Surface plasmon resonance, Springer Ser Chem Sens Biosens, vol 4. Springer, Berlin Heidelberg New York (in this volume)
64. Löfås S, Mcwhirter A (2006) The art of immobilization for SPR sensors. In: Homola J (ed) Surface plasmon resonance, Springer Ser Chem Sens Biosens, vol 4. Springer, Berlin Heidelberg New York (in this volume)

Springer Ser Chem Sens Biosens (2006) 4: 191–206
DOI 10.1007/5346_020

Published online: 8 July 2006

SPR Biosensors for Environmental Monitoring

Jakub Dostálek · Jiří Homola (✉)

Institute of Radio Engineering and Electronics, Prague, Czech Republic
homola@ure.cas.cz

Keywords Biosensor · Dioxins · Environmental monitoring · Heavy metals · Inorganic contaminants · Microbial pathogens · Organic contaminants · Pesticides · Phenols · Polycyclic aromatic hydrocarbons · Polychlorinated biphenyls · Sensor · Surface plasmon resonance · Toxins

1 Introduction

An increasing number of chemical and biological substances are released into the environment every year as a result of industrial and agricultural activity. Numerous substances have been identified as harmful and subjected to regulatory measures (e.g., polycyclic aromatic hydrocarbons, pesticides, heavy metals, polychlorinated biphenyls, dioxins), risks to human and wildlife organisms of others are still being assessed (e.g., surfactants, pharmaceuticals, and nanoparticles) [1–6]. Harmful substances can be discharged into various environments including air, soil, and water through which they can interfere with human and wildlife organisms as endocrine disrupters, carcinogens, or (geno)toxicants. In order to protect public health and the local ecosystems from the harmful effects of these compounds, efficient tools for their rapid detection are urgently needed.

Currently, detection of harmful contaminants is performed using established analytical techniques, such as gas chromatography (GC), liquid chromatography (LC) and mass spectrometry (MS) (detection of small organic pollutants [7, 8]), culturing and polymerase chain reaction (PCR) combined with DNA arrays (detection of microbial pathogens [9]), and immunoassays (detection of small organic pollutants and microbial pathogens [10, 11]). However, these methods require sophisticated equipment and laborious sample preparation and are therefore performed by highly trained personnel in specialized laboratories.

Chemical sensors and biosensors for environmental monitoring present an interesting alternative to these conventional methods and offer numerous attractive features such as ease of use, low cost, portability, and the ability to perform detection in the field [12–16]. Optical biosensors based on surface plasmon resonance (SPR) show potential for rapid and sensitive detection of chemical and biological contaminants in the environment. SPR sensors provide a generic platform which, in conjunction with appropriate biorecognition elements, can be tailored for detection of numerous compounds related to environmental protection [17–19].

This chapter reviews applications of SPR biosensors for detection of chemical and biological contaminants that present environmental risks, including organic chemicals, inorganic chemicals, microbial pathogens, and toxins.

2 Organic Contaminants

Organic contaminants that present a concern to environmental protection include pesticides (used in agriculture), polycyclic aromatic hydrocarbons (PAHs, a by-product of incomplete combustion), polychlorinated biphenyls (PCBs, components of coolants and lubricants), phenols (used in the production of plastics and pesticides), dioxins (unwanted by-products of many industrial processes including incineration and chemical manufacturing of phenols, PCBs, and herbicides) and alkyphenols (surfactants in agrochemicals and household cleaning products).

2.1 Pesticides

In agriculture, several hundreds of different pesticides have been used worldwide over the last few decades. Owing to their widespread applications and persistence in the environment, pesticides are accumulating in media such as soil and ground water. Many pesticides exhibit endocrine-disrupting activity, which poses a threat to public health and local ecosystems, and are therefore regulated. In the European Union, the maximum allowable concentrations in drinking water for individual pesticides and pesticides in total are 0.1 ng mL^{-1} and 0.5 ng mL^{-1}, respectively [20]. In the United States, for drinking water the US Environmental Protection Agency (EPA) sets the maximum allowed concentrations for the most common pesticides such as atrazine and simazine to 3 ng mL^{-1} and 4 ng mL^{-1}, respectively [21].

As pesticides are rather small molecules (e.g., the molecular weight of atrazine is 216 Da), their detection is usually performed using the inhibition assay (see Chap. 7 in this volume [54]). In this type of sensor, the analyzed sample is pre-incubated with an antibody for the targeted analyte. Subsequently, the mixture is injected into the SPR sensor with an analyte derivative immobilized on the sensor surface, and the binding of the unreacted antibody to the analyte derivative is measured. The presence of the targeted analyte in the sample is detected as a decrease of antibody binding to the analyte derivative. Figure 1 shows a typical sensorgram and calibration curve for an inhibition assay-based SPR biosensor for detection of pesticides developed at the Institute of Radio Engineering and Electronics, Prague.

The first SPR immunosensor for detection of pesticides was developed by Minunni et al. [22] in the early 1990s. They used an SPR sensor developed by Biacore AB, Sweden, with the atrazine derivative bound to dextran matrix on the sensor chip. The detection of atrazine was performed using the inhibition assay and monoclonal antibodies. The sensor response was subsequently amplified by secondary antibody, which was bound to the antibody captured by the atrazine derivative (sandwich assay, see Chap. 7 in this volume [54]). This biosensor was demonstrated to measure atrazine in distilled and tap water within the range 0.05–1 ng mL^{-1} in 15 min and exhibited relatively low cross-reactivity with simazine and tetrabutyl atrazine (20%). The sensor surface was regenerated with 100 mM sodium hydroxide in 20% acetonitrile.

Mouvet et al. developed an integrated optical (IO) SPR sensor for simazine [23]. A triazine derivative was immobilized on the sensor using the

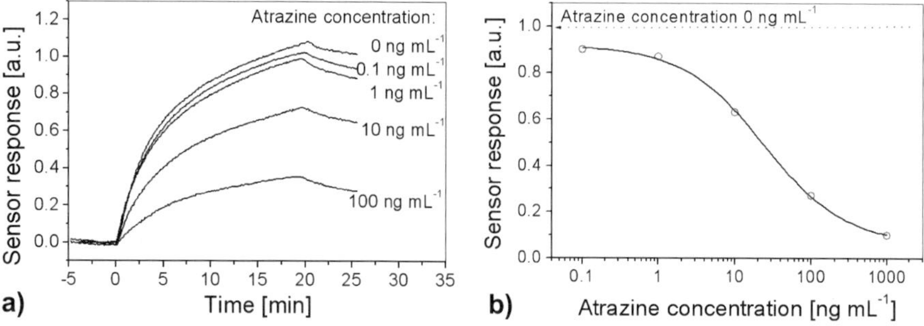

Fig. 1 Inhibition assay for detection of atrazine: **a** SPR sensorgram obtained while flowing over the sensor surface aqueous samples with atrazine at the concentrations 0, 0.1, 1, 10, and 100 ng mL^{-1} incubated with antibody; **b** calibration curve of the sensor

dextran chemistry. Using inhibition assay this system allowed detection of simazine in the range 0.11–1.1 ng mL^{-1} within 20 min. The sensor surface was shown to be regenerable allowing up to 200 detection cycles on a single chip. The regeneration was performed by the sequential incubation in pepsin (2 mg mL^{-1}) and solution of 50% acetonitrile and 1% proprionic acid. The sensor exhibited relatively high cross reactivity with atrazine and tetrabutyl atrazine (approximately 60%). Detection in natural surface and ground water samples without any sample preparation other than sample filtration was demonstrated. Harris et al. [24] combined this sensor with two types of antibody receptors (IgG and their Fab fragments) for detection of simazine; calibration curves achieved for these two biorecognition elements are shown in Fig. 2.

Another widely used pesticide, 2-4 dichlorophenoxyacetic acid (2,4-D), was detected with an SPR immunosensor by Gobi et al. [25]. They used inhibition assay and a sensor chip on which the conjugate of BSA and 2,4-D derivative was physisorbed. The detection was performed with monoclonal antibodies in phosphate buffer saline (PBS) using a commercial SPR-20 instrument (from DKK-TOA, Japan). A detection range of 0.5 ng mL^{-1} to 1 μg mL^{-1} and a detection time of 20 min were achieved. Regeneration of the sensor for up to 20 detection cycles was performed using pepsin (10 μg mL^{-1}).

Nakamura et al. demonstrated direct detection of herbicides by using a heavy-subunit-histidine-tagged photosynthesis reaction center (HHisRC) from bacterium *Rhodobacter sphaerodies* [26] (Fig. 3). They used a Biacore X instrument (from Biacore AB, Sweden) and a sensor chip with dextran matrix

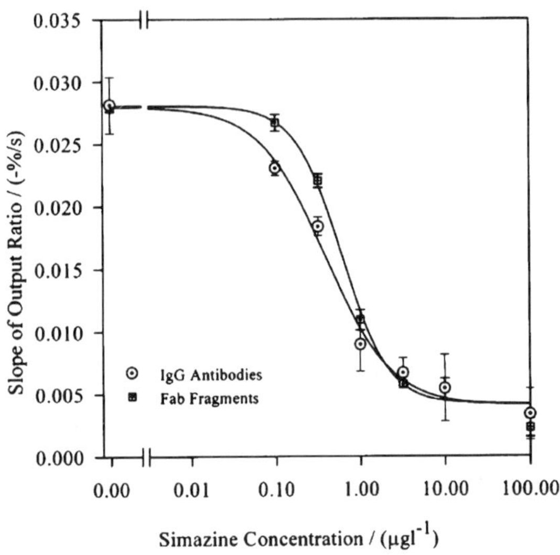

Fig. 2 Calibration curve of integrated optical SPR immunosensor for detection of simazine using inhibition assay and IgG antibodies and Fab fragments [24]

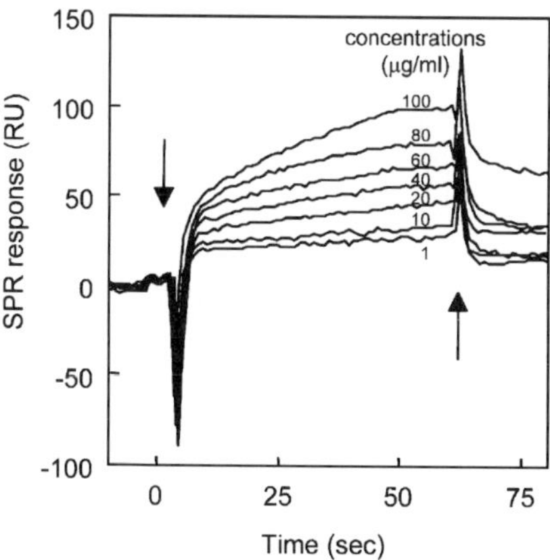

Fig. 3 Sensorgrams obtained from direct atrazine SPR biosensor for atrazine binding to the heavy-subunit-histidine-tagged photosynthesis reaction center (HHisRC) immobilized on the sensor chip; atrazine concentrations 1–100 $\mu g\,mL^{-1}$ [26]

into which HHisRC was immobilized by nickel chelation chemistry. Detection of atrazine in buffer in the concentration range 1–100 $\mu g\,mL^{-1}$ was demonstrated. Detection time in the order of minutes was achieved.

Chegel et al. reported an SPR biosensor based on displacement of plastoquione from D1 protein interacting with photosynthesis-inhibiting pesticides [27]. They used an SPR biosensor system with angular modulation of SPR and D1 protein attached to the sensor chip via phisisorption on thiol self-assembled monolayer. When exposed to atrazine, plastoquione was displaced from D1 protein producing a change in the SPR signal. This sensor allowed detection of atrazine within the concentration range 50–5000 $ng\,mL^{-1}$.

Detection of atrazine based on specifically expressed mRNA in *Saccharomyces cerevisiae* bacteria exposed to atrazine was reported by Lim et al. [28]. The cells were brought into contact with the analyzed sample, disrupted, and the amount of expressed P450 mRNA was measured using an SPR biosensor Biacore 2000 (Biacore AB, Sweden) with complementary oligonucleotide probes. These probes were biotin-labeled and immobilized on the sensor surface using streptavidin–biotin chemistry. Detection of atrazine in the range 1 $pg\,mL^{-1}$ to 1 $\mu g\,mL^{-1}$ was reported. The analysis, including bacteria incubation, disruption, and mRNA detection, was completed in 15 min.

2.2 Polycyclic Aromatic Hydrocarbons

Polycyclic aromatic hydrocarbons (PAH) are a group of over 100 different chemicals that are formed during the incomplete combustion of coal, oil, gas, garbage, or other organic substances and can be found in air, wa-

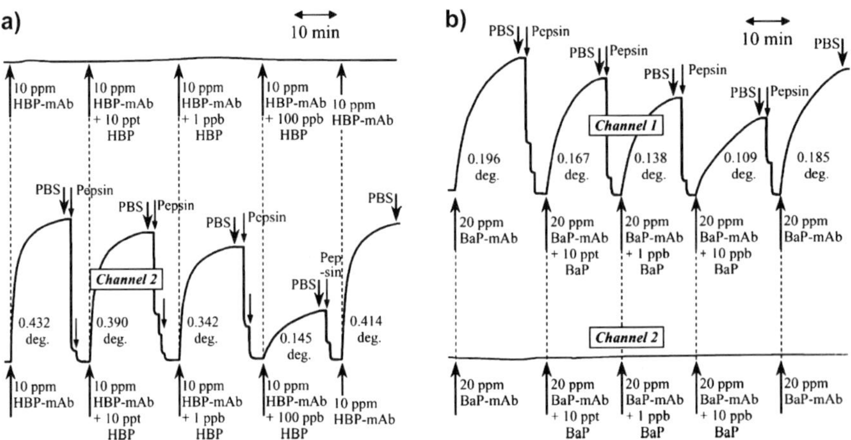

Fig. 4 Sensorgram obtained for different concentrations of benzo(a)pyrene (BaP) and 2-hydroxybiphenyl (HBP) detected in buffer by inhibition assay. Channel 1 and channel 2 are modified with BSA–BaP and BSA–HBP conjugates, respectively [29]

ter, and sediments. PAHs are regulated due to their endocrine-disrupting and carcinogenic activity (e.g., US EPA sets a maximum concentration for benzo[a]pyrene in drinking water at 0.2 ng mL^{-1} [21]).

A SPR immunosensor for detection of benzo(a)pyrene (BaP) and 2-hydroxybiphenyl (HBP, metabolite of BaP) by an inhibition assay was reported by Gobi et al. and Miura et al. [29–31]. They immobilized BaP and HBP conjugates of bovine serum albumin (BSA) by physical adsorption on the SPR sensor chip, which was used in a two-channel SPR-20 sensor (DKK, Japan). Using monoclonal antibodies against BaP and HBP, simultaneous detection of BaP and HBP in buffer was demonstrated with a detection limit as low as 0.01 ng mL^{-1} [29]. The detection was performed in 15 min and the sensor was regenerated for repeated measurements by using pepsin and a pH change. The sensor exhibited negligible cross-sensitivity between BaP and HBP (Fig. 4).

2.3 Polychlorinated Biphenyls

Polychlorinated biphenyls (PCB) – currently banned compounds – were formerly used in hydraulic fluids, plasticizers, adhesives, fire retardants, and pesticide extenders. These contaminants are persistent in the environment and are present in sediments at the bottom of lakes, rivers, and seas. As they exhibit carcinogenic and endocrine-disrupting activity, they are subject to regulation. For instance, in the United States, the maximum allowed concentration of PCBs in drinking water is 0.5 ng mL^{-1} [21].

Shimomura et al. used a Biacore 2000 SPR sensor instrument (Biacore AB, Sweden) for detection of PCB 3,3′,4,4′,5-pentachlorobiphenyl [32]. They employed competition assay format and the sensor chip with polyclonal antibodies immobilized in the dextran matrix. The sample was mixed with a conjugate of PCB-horseradish peroxidase (HRP) and injected into the sensor. The presence of the analyte was detected as a decrease in binding of PCB–HRP conjugate. The detection was performed in 15 min with a detection limit of 2.5 ng mL^{-1} in buffer. The sensor was demonstrated to be regenerable by 0.1 M hydrochloric acid.

2.4 Phenolic Contaminants

Phenolic compounds relevant to environmental protection include bisphenol A, nonylhenol, 2,4-dichlorophenol, phenol, hydroquinone, resorcinol, phloroglucinol, and catechol.

Bisphenol A (BPA) is a compound which is currently not regulated, but as it exhibits weak estrogenic properties it is a suspected endocrine disrupter [33]. BPA is widely used as a plasticizer in plastics such as polycarbonate and epoxy resins and thus it is a concern for water quality. Soh et al. developed an SPR immunosensor based on inhibition assay to detect BPA [34]. They used a SPR-20 sensor instrument (DKK, Japan) with the sensor chip modified with thiol monolayer on which BPA was immobilized through BPA succinimidyl ester. Using a monoclonal antibody, detection of BPA in buffer at concentrations as low as 10 ng mL^{-1} was achieved. Detection time was approximately 30 min and the sensor was demonstrated to be regenerable using 0.01 M hydrochloric acid.

Large amounts of polyethoxylated alkyphenol detergents are released into the environment, where they degrade forming more toxic and dangerous compounds such as alkyphenols. Alkylphenols are not yet regulated, however, for instance US EPA has recommended a maximum 1-h average concentration (acute criterion) and maximum 4-day average concentration (chronic criterion) of 27.9 and 5.9 ng mL^{-1}, respectively, for nonylphenol in freshwater [35]. Recently, Samsonova et al. reported an inhibition assay-based SPR biosensor for detection of 4-nonylphenol [36]. They used a Biacore Q device (from Biacore AB, Sweden) and a sensor chip with dextran matrix on which 9-(*p*-hydroxyphenyl)nonanoic acid was immobilized using amine coupling chemistry. Using monoclonal antibodies, a detection limit of 2 ng mL^{-1} in buffer was achieved. The detection was performed in less than 3 min including a 30-s regeneration step. The sensor was regenerated by 100 mM sodium hydroxide in 10% acetonitrile. Furthermore, the sensor was applied for detection of 4-nonyphenol in shellfish with a detection limit of 10 ng g^{-1} (Fig. 5) (time for sample preparation was approximately 1 h).

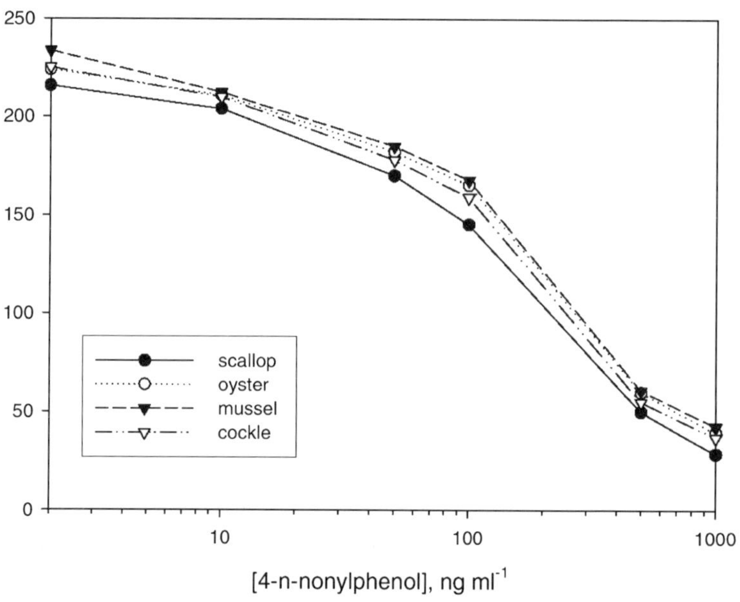

Fig. 5 Calibration curve for SPR sensor-based detection of 4-nonylphenol in four different shellfish samples by inhibition assay [36]

2,4-Dichlorophenol is the major dioxin precursor. Currently it is not regulated, but it is listed among the drinking water contaminant candidates which the US EPA intends to regulate in future [37]. Soh et al. developed an SPR biosensor for detection of this compound based on a competition assay [38]. They used a SPR-20 sensor instrument (DKK, Japan) and the sensor chip with monoclonal antibodies against 2,4-dichlorophenol immobilized on the sensor surface via gold-binding peptide and protein G. The assay was based on the competition between the binding of analyte present in a sample and added conjugate of BSA–2,4-dichlorophenol. Detection of 2,4-dichlorophenol in buffer with a detection limit of 20 ng mL^{-1} was demonstrated.

Write et al. explored direct detection of various phenolic compounds (phenol, hydroquinone, resorcinol, phloroglucinol, and catechol) using an SPR sensor with intensity modulation and synthetic receptors loaded in a polymer or sol-gel layer [39]. Their preliminary experiments, consisting of the detection of phenols in buffer, suggested that detection of phenols at millimolar concentration levels ($\sim 100\,\mu g\,mL^{-1}$) is feasible using this approach.

Detection of phenols based on toxicity measurement was carried out by Chio et al. [40]. In their experiments, they used a Multiscope SPR sensor (Optrel, Germany) and *Escherichia coli* cells immobilized on an SPR sensor chip via synthetic cystein-terminated oligopeptides. When the immobilized cells were exposed to phenol, a decrease in the SPR signal was observed due to the damage of the cells. Using this approach, phenol in buffer was detected at concentrations down to 5 μg mL^{-1}.

2.5 Dioxins

Dioxins are released in the environment in emissions from the incineration of municipal refuse and certain chemical wastes and in exhaust from automobiles powered by leaded gasoline. Dioxins are highly persistent and accumulate in the environment. They are highly toxic and exhibit endocrine-disrupting activity. Therefore, they are regulated by authorities, e.g., in the United States for the most toxic 2,3,7,8-TCDD the maximum allowed concentration in drinking water is 10^{-4} ng mL^{-1} [21].

Shimomura et al. developed an SPR biosensor for detection of 2,3,7,8-TCDD [32]. They used a Biacore 2000 instrument (Biacore AB, Sweden) and competition assay with monoclonal antibodies against 2,3,7,8-TCDD. The antibody was immobilized in the dextran matrix on the sensor chip by amine coupling chemistry. The sample was mixed with a conjugate of 2,3,7,8-TCDD–horseradish peroxidase (HRP) and injected into the sensor. The assay was completed in 15 min. A detection limit of 0.1 ng mL^{-1} was achieved and the sensor was shown to be regenerable using 0.1 M hydrochloric acid.

2.6 Trinitrotoluene

There is a demand for rapid detection of explosives, especially for environmental restoration and humanitarian demining. Among these compounds, trinitrotoluene (TNT) has attracted a great deal of attention as a main constituent of most of antipersonnel landmines and due to its toxic, mutagenic, and carcinogenic effects.

Strong et al. developed an SPR biosensor for detection of TNT [41]. They used a SPREETA sensor (from Texas Instruments, USA) with disposable sensor chip coated with BSA–trinitrobenzen conjugate. The detection of TNT was performed by inhibition assay for TNT concentrations down to 1 μg g^{-1} for soil samples. The time needed for sample preparation (suspension followed by centrifugation) and analysis of the liquid supernatant were approximately 10 and 6 min, respectively. The sensor was shown to be regenerable using a solution with 0.1 M sodium chloride and 0.1% Triton X-100.

Sandakaran et al. reported another SPR immunosensor for TNT. For inhibition assay-based detection, BSA–2,4,6 trinitrophenol conjugate was anchored to the sensor surface by physical sorption. The detection was performed using the sensor instrument SPR-670 (Nippon Laser and Electronics, Japan). With polyclonal antibodies against BSA–TNP, a detection limit as low as 0.09 ng mL^{-1} was achieved in buffer [42]. The sensor exhibited very low cross-sensitivity to other similar compounds such as 1,4-DNT, 1,3-DNB, 2A-4,6-DNT, and 4A-4,6-DNT. The sensor was demonstrated to be regenerable and a single detection cycle was performed in 22 min. The regeneration was performed by pepsin.

3 Inorganic Contaminants

Heavy metals belong to the group of harmful inorganic contaminants. They are released in the environment from factories and coal-burning power plants and do not naturally decompose in the environment [43]. Currently, they are regulated and, for instance, in freshwater US EPA sets maximum allowed concentrations for heavy metals such as cadmium (chronic criterion 0.15 ng mL^{-1} [44]), copper (chronic criterion 9 ng mL^{-1} [45]), nickel (chronic criterion 52 ng mL^{-1} [46]) and zinc (chronic criterion 120 ng mL^{-1} [47]).

Detection of heavy metals was demonstrated by Wu et al. who used a Biacore X instrument (Biacore AB, Sweden) with rabbit metallothinein coupled to dextran matrix on the sensor chip [48]. Metallothinein is a protein that can be found in the cells of many organisms and is known to bind to metals (especially cadmium and zinc). Model experiments in which metallothein was used as a receptor demonstrated the potential of this sensor to directly detect Cd, Zn, and Ni in buffer at concentrations down to 0.1 μg mL^{-1}.

Another approach to direct detection of Cu^{2+} ions was reported by Ock et al. [49]. They used a sensor based on attenuated total reflection and angular modulation of SPR with squarylium dye (SQ) loaded in a thin polymer layer as a recognition element. This dye changes its absorption when it in-

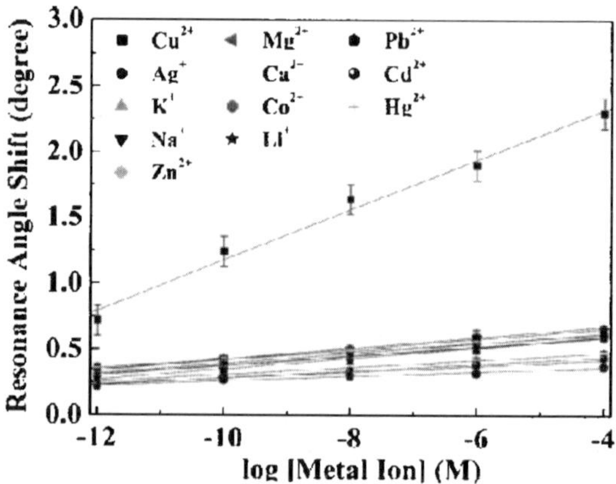

Fig. 6 Calibration curve of direct Cu^{2+} SPR sensor with an SQ dye loaded in a polymer layer deposited on the surface [49]

teracts with Cu^{2+} ions. Owing to anomalous dispersion accompanying this absorption, enhanced refractive index changes can be observed when SQ dye is exposed to Cu^{2+} ions. By tuning the operating wavelength of the SPR sensor to the absorption wavelength of the dye, selective detection of Cu^{2+} in the concentration range 1×10^{-12}–1×10^{-4} M (0.063 pg mL^{-1} to 6.3 μg mL^{-1}) was demonstrated (Fig. 6).

4
Microbial Pathogens and Toxins

Besides anthropogenic contaminants, other harmful compounds such as bacterial pathogens and toxins can also be found in the environment. For instance, bacterial pathogens including *Legionella pneumophila*, *Salmonella typhimurium*, *Yersenia enterocolitica* and *Escherichia coli* O157:H7 are known to be able to outbreak via potable water systems, cooling towers, or heat exchanger systems. These bacteria cause significant threat to populations as they can cause various illnesses such as Legionnaire's disease (*Legionella pneumophila*), gastroenteritis (*Salmonella typhimurium*) and diarrhea (*Yersinia enterocolitica*, *Escherichia coli* O157:H7). In addition, various toxins can be released into the environment from natural sources such as algae. Among these, domoic acid (DA, a low molecular weight toxin produced by the marine diatom *Pseudonitzshia pungens*) is identified as posing a risk to human populations as it can accumulate in edible shellfish and exhibits neurotoxic effects.

Recently, Oh et al. have developed an SPR immunosensor capable of simultaneous detection of *Legionella pneumophila*, *Salmonella typhimurium*, *Yersinia enterocolitica* and *Escherichia coli* O157:H7 [50, 51]. Monoclonal antibodies against these bacteria were immobilized in individual sensing channels of the sensor chip of a Multiscope SPR sensor (Optrel, Germany) via protein G. For bacteria in buffer, the detection limit of 10^5 cfu mL^{-1} and good specificity were achieved. No special treatment of bacteria was performed.

Lotierzo et al. developed a photografted molecular imprinted polymer-based SPR sensor for detection of domoic acid [52]. They used a Biacore 3000 sensor instrument (Biacore AB, Sweden) and the competition assay. The conjugate of horseradish peroxidase and DA was added into analyzed sample and the competition of binding of DA and the conjugate to the sensor chip was measured. As follows from the calibration curve (Fig. 7), a detection limit as low as 5 ng mL^{-1} was achieved. Recently, Yu et al. detected DA using a two-channel sensor based on attenuated total reflection and wavelength modulation of SPR [53]. DA was immobilized on the sensor chip by thiol chemistry, and inhibition assay involving monoclonal antibodies was employed. Detection of DA at concentrations as low as 0.1 ng mL^{-1} was achieved. The sensor was regenerated for up to 20 detection cycles using 50 mM hy-

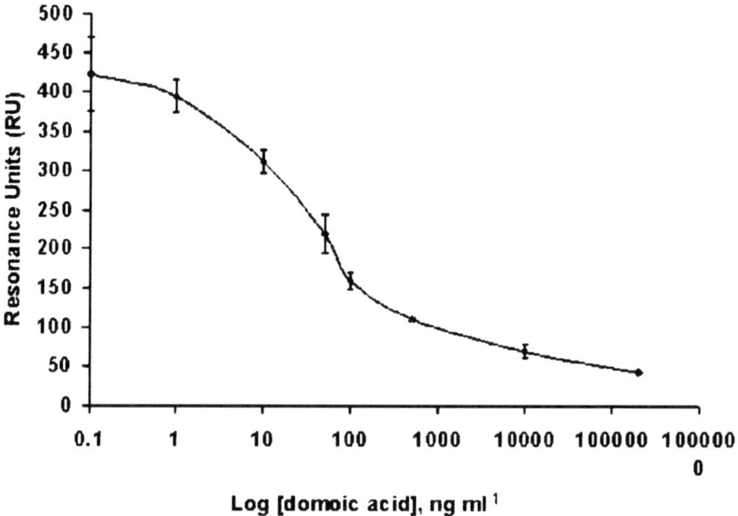

Fig. 7 Calibration curve of SPR sensor with photografted molecular imprinted polymer and competition assay [52]

droxide solution and demonstrated an ability to maintain its activity even after two months of storage.

5 Summary and Outlook

In recent years, we have witnessed an increasing research and development activity aiming at introducing SPR biosensor technology to environmental monitoring. Numerous substances of environmental concern have been targeted and SPR biosensors for their rapid detection at practically relevant concentrations have been demonstrated (Table 1). However, most of the reported biosensors for environmental contaminants have taken advantage of commercially available laboratory SPR sensor instruments not suitable for the use in the field and dealt with rather pure samples (Table 1). In the future, we expect that advances in the development of mobile SPR sensor platforms, immobilization methods, and biorecognition elements will enable rapid detection and continuous monitoring of environmental contaminants in the field and thus substantially contribute to the protection of local ecosystems and public health.

Acknowledgements This work was supported by grants from the United States Food and Drug Administration (contract FD-U-002250) and the European Commission (contract QLK4-CT-2002-02323).

Table 1 Overview of environmental contaminants detected with SPR biosensors

Compound	Sensor system	Detection limit	Detection time (min)	Biorecognition element and assay format	Detection matrix	Refs.
Pesticides:						
Atrazine	Biacore	0.05 ng mL^{-1}	15	Antibody, inhibition	Drinking water	[22]
	Biacore 2000	1 pg mL^{-1}	15	mRNA expression	Buffer	[28]
	Home built (ATR)	50 ng mL^{-1}	Not available	D1 protein, displacement	Buffer	[27]
Simazine	Home built (IO chip)	0.11 ng mL^{-1}	20	Antibody, inhibition	Surface and ground water	[24]
2,4-Dichloro-phenoxy acetic acid	SPR-20	0.5 ng mL^{-1}	20	Antibody, inhibition	Buffer	[25]
Herbicides (atrazine)	Biacore X	1 μg mL^{-1}	< 10	Photosynthesis RC, direct	Buffer	[26]
Phenolic compounds:						
Bisphenol A	SPR-20	10 ng mL^{-1}	Not available	Antibody, inhibition	Buffer	[34]
4-Nonylphenol	Biacore Q	2 ng mL^{-1}/ 10 ng g^{-1}	2.5/60	Antibody, inhibition	Buffer/ Shellfish	[36]
Phenol	Multiscope	5 μg mL^{-1}	Not available	Whole cell, direct	Buffer	[40]
Phenols	Home built (ATR)	1 mM (∼ 100 μg mL^{-1})	Not available	Synthetic receptors, direct	Purified water	[39]
2,4-Dichlorophenol	SPR-20	20 ng mL^{-1}	Not available	Antibody, inhibition	Buffer	[38]
Other organic compounds:						
3,3′,4,4′,5-Pentachloro-biphenyl	Biacore 2000	2.5 ng mL^{-1}	15	Antibody, competition	Buffer	[32]
2-Hydroxybiphenyl	SPR-20	0.01 ng mL^{-1}	15	Antibody, inhibition	Buffer	[29, 30]
2,3,7,8-Tetra-chloro-dibenzo-*p*-dioxin	SPR-20	0.1 ng mL^{-1}	15	Antibody, competition	Buffer	[32]

Table 1 (continued)

Compound	Sensor system	Detection limit	Detection time (min)	Biorecognition element and assay format	Detection matrix	Refs.
Benzo[a]pyrene	SPR-20	0.010 ng mL^{-1}	15	Antibody, inhibition	Buffer	[29, 30]
Trinitrotoluene	SPR 670	0.09 ng mL^{-1}	22	Antibody, inhibition	Buffer	[42]
Trinitrotoluene	SPREETA	1 μg g^{-1}	16	Antibody, inhibition	Soil	[41]
Heavy metals:						
Cd, Zn, and Ni	Biacore X	$\sim 10^{-6}$ M ($\sim$ 0.1 μg mL^{-1})	Not available	Metallothinein, direct	Buffer	[48]
Cu^{2+}	Home built (ATR)	$\sim 10^{-12}$ M (0.063 pg mL^{-1})	Not available	Squadrylium dye, direct	Buffer	[49]
Microbial analytes and toxins:						
Escherichia coli	Multiscope	10^5 cfu mL^{-1}	Not available	Antibody, direct	Buffer	[51]
Legionella pneumophila	Multiscope	10^5 cfu mL^{-1}	Not available	Antibody, direct	Buffer	[50, 51]
Salmonella typhimurium	Multiscope	10^5 cfu mL^{-1}	Not available	Antibody, direct	Buffer	[51]
Yersenia enterocolitica	Multiscope	10^5 cfu mL^{-1}	Not available	Antibody, direct	Buffer	[51]
Domoic acid	Home built (ATR)	0.1 ng mL^{-1}	60	Antibody, inhibition	Buffer	[53]
Domoic acid	Biacore 3000	5 ng mL^{-1}	16	Molecular imprinted polymer matrix, competition	Buffer	[52]

References

1. Velasquez IB, Jacinto GS, Valera FS (2002) Marine Pollut Bull 45:210
2. Daughton C (2004) Environ Impact Assess Rev 24:711
3. Petrovic M, Eljarrat E, de Alda MJL, Barcelo D (2004) Anal Bioanal Chem 378:549
4. Tanabe S (2002) Marine Pollut Bull 45:69
5. Allera A, Lo S, King I, Steglich F, Klingmuller D (2004) Toxicology 205:75
6. Lemaire G, Terouanne B, Mauvais P, Michel S, Rahmani R (2004) Toxicol Appl Pharmacol 196:235
7. Petrovic M, Hernando MD, Diaz-Cruz MS, Barcelo D (2005) J Chromatogr A 1067:1
8. Ferrer I, Thurman EM (2003) Trends Anal Chem 22:750
9. Simpson JM, Domingo JWS, Reasoner DJ (2002) Environ Sci Technol 36:5279
10. Hock B, Dankwardt A, Kramer K, Marx A (1995) Anal Chim Acta 311:393
11. Mouvet C, Broussard S, Jeannot R, Maciag C, Abuknesha R, Ismail G (1995) Anal Chim Acta 311:331
12. Kroger S, Piletsky S, Turner APF (2002) Marine Pollut Bull 45:24
13. Rogers KR, Lin JN (1992) Biosens Bioelectron 7:317
14. Rogers KR, Mascini M (1998) Field Anal Chem Technol 2:317
15. Baeumner A (2003) Anal Bioanal Chem 377:434
16. Rodriguez-Mozaz S, de Alda M, Marco M, Barceló D (2005) Talanta 65:291
17. Homola J (2003) Anal Bioanal Chem 377:528
18. Rich RL, Myszka DG (2000) Curr Opin Biotechnol 11:54
19. Homola J, Yee SS, Gauglitz G (1999) Sensor Actuator B Chem 54:3
20. Council Directive 98/83/EC of 3 November 1998 on the quality of water intended for human consumption. J Eur Commun L 330/32
21. US national primary drinking water regulations, 40CFR141 – Part 141
22. Minunni M, Mascini M (1993) Anal Lett 26:1441
23. Mouvet C, Harris R, Maciag C, Luff B, Wilkinson J, Piehler J, Brecht A, Gauglitz G, Abuknesha R, Ismail G (1997) Anal Chim Acta 338:109
24. Harris R, Luff B, Wilkinson J, Piehler J, Brecht A, Gauglitz G, Abuknesha R (1999) Biosens Bioelectron 14:377
25. Gobi KV, Tanaka H, Shoyama Y, Miura N (2005) Sensor Actuator B Chem 111–112:562
26. Nakamura C, Hasegawa M, Nakamura N, Miyake J (2003) Biosens Bioelectron 18:599
27. Chegel V, Shirshov Y, Piletskaya E, Piletsky S (1998) Sensor Actuator B Chem 48:456
28. Lim T, Oyama M, Ikebukuro K, Karube I (2000) Anal Chem 72:2856
29. Gobi K, Miura N (2004) Sensor Actuator B Chem 103:265
30. Gobi K, Tanaka H, Shoyama Y, Miura N (2004) Biosens Bioelectron 20:350
31. Miura N, Sasaki M, Gobi KV, Kataoka C, Shoyama Y (2003) Biosens Bioelectron 18:953
32. Shimomura M, Nomura Y, Zhang W, Sakino M, Lee K, Ikebukuro K, Karube I (2001) Anal Chim Acta 434:223
33. Staples CA, Dorn PB, Klecka GM, Block ST, Harris LR (1998) Chemosphere 36:2149
34. Soh N, Watanabe T, Asano Y, Imato T (2003) Sensor Mater 15:423
35. Ambient aquatic life water quality criteria for nonylphenol – draft, EPA-822-R-03-029
36. Samsonova JV, Uskova NA, Andresyuk AN, Franek M, Elliott CT (2004) Chemosphere 57:975
37. Fact sheet: the drinking water contaminant candidate list – the source of priority contaminants for the drinking water EPA 815-F-05-001
38. Soh N, Tokuda T, Watanabe T, Mishima K, Imato T, Masadome T, Asano Y, Okutani S, Niwa O, Brown S (2003) Talanta 60:733

39. Wright J, Oliver J, Nolte R, Holder S, Sommerdijk N, Nikitin P (1998) Sensor Actuator B Chem 51:305
40. Choi JW, Park KW, Lee DB, Lee W, Lee WH (2005) Biosens Bioelectron 20:2300
41. Strong A, Stimpson DI, Bartholomew DU, Jenkins TF, Elkind JL (1999) SPIE 3710:362
42. Shankaran D, Gobi K, Sakai T, Matsumoto K, Toko K, Miura N (2005) Biosens Bioelectron 20:1750
43. Sharma RK, Agrawal M (2005) J Environ Biol 26:301
44. 2001 Update of ambient water quality criteria for cadmium, EPA-822-R-01-001
45. Ambient water quality criteria for copper EPA 440/5-84-031
46. Ambient water quality criteria for nickel EPA 440/5-80-060
47. Ambient water quality criteria for zinc EPA 440/5-80-079
48. Wu C, Lin L (2004) Biosens Bioelectron 20:864
49. Ock K, Jang G, Roh Y, Kim S, Kim J, Koh K (2001) Microchem J 70:301
50. Oh B, Kim Y, Lee W, Bae Y, Lee W, Choi J (2003) Biosens Bioelectron 18:605
51. Oh BK, Lee W, Chun BS, Bae YM, Lee WH, Choi JW (2005) Biosens Bioelectron 20:1847
52. Lotierzo M, Henry O, Piletsky S, Tothill I, Cullen D, Kania M, Hock B, Turner A (2004) Biosens Bioelectrons 20:145
53. Yu Q, Chen S, Taylor A, Homola J, Hock B, Jiang S (2005) Sensor Actuator B Chem 107:193
54. Dostálek J, Ladd J, Taylor A, Jiang S, Homola J (2006) Detection of Biological and Chemical Analytes. In: Homola J (ed) Surface Plasmon Resonance Based Sensors. Springer Ser Chem Sens Biosens, vol 4. Springer, Berlin Heidelberg New York (in this volume)

Springer Ser Chem Sens Biosens (2006) 4: 207–227
DOI 10.1007/5346_021

Published online: 8 July 2006

SPR Biosensors for Food Safety

Jon Ladd · Allen Taylor · Shaoyi Jiang (✉)

Department of Chemical Engineering, University of Washington, Box 351750, Seattle, WA 98195-1750, USA
sjiang@u.washington.edu

Keywords Bacteria · Biosensor · Food safety · Foodborne pathogens · Sensor · Surface plasmon resonance · Toxins

1 Introduction

Detection of foodborne diseases and food contaminants is an important application for biosensor technology development. More than 200 known diseases are transmitted through food. In the United States alone, foodborne diseases cause an estimated 76 million illnesses and 325 000 hospitalizations a year [1]. Foodborne transmission accounts for approximately 36% of the total number of reported foodborne illnesses. The majority of illnesses attributed to foodborne transmission are caused by viruses (67%). Bacteria

(30%) and parasites (3%) account for the remaining cases. Known foodborne pathogens account for an estimated 14 million illnesses, 60 000 hospitalizations, and 1800 deaths a year in the United States. Almost 90% of these deaths can be attributed to five pathogens: *Salmonella, Listeria, Toxoplasma, Campylobacter*, and *Escherichia coli* O157:H7.

The ability to identify contaminated food samples is of great importance to the food processing industry as well as to regulatory agencies. There is a need for accurate techniques to rapidly detect the presence of foodborne pathogens. Conventional techniques for analyzing food for the presence of pathogens include culturing methods [2], polymerase chain reaction (PCR) [3], flow cytometry [4], and enzyme-linked immunosorbent assay (ELISA) [5]. Problems faced by these conventional techniques include the need for highly specialized laboratory equipment and training, as well as laborious pretreatment of samples. Thus, portable, simple-to-use techniques are more suited for use in the food industry. Surface plasmon resonance (SPR) is one such technique that is being applied to the food industry. Biosensors based on the SPR phenomenon offer significant advantages over other commonly used sensing technologies. SPR biosensors are capable of performing real-time detections, making on-line monitoring of food processing possible. In addition to real-time detections, SPR biosensors are also able to produce quantitative and sensitive detections in less than an hour. These real-time, quantitative detections are done without the necessity of labeled (e.g. fluorescence) compounds.

To establish SPR biosensors as a practical technology in the food safety industry, an economically viable and easy-to-use approach is needed for detecting analytes in food samples. To achieve this goal, sensors must be robust, requiring little-to-no maintenance. For food pathogens, low limits of detections vary based on the infective dose of the pathogen being detected. This can range from ten cells for *E. coli* O157:H7 [6] to 1.1 ng/kg body weight in the case of Botulinum toxins [7]. The specificity of the sensing platform is also extremely important. False positive alarms and false negatives must be kept to a minimum. A recent call for proposals from the United States Department of Homeland Security required a false positive rate of ≤ 1 in 1 000 000 [8]. All of these requirements must be carried out on analytes residing in real-world samples, such as ground beef, apple juice, etc.

In this review, foodborne pathogens will be divided and discussed on a basis of the size of the analyte being detected. This division separates pathogens into categories posing different challenges for detection: bacteria, proteins, and low molecular weight compounds. For bacteria, which are on the order of a micron in size, the biggest hindrance for sensitive detections is the slow rate of diffusion to the sensing surface. Pathogenic proteins are probably the best suited of the three classes of analytes for detection with SPR biosensors. They typically range in molecular weight from 5 kDa to 150 kDa, giving them both a high rate of diffusion to the sensing surface and enough mass to produce a measurable signal at higher concentrations. The detection of low molecular weight compounds produces a different set of challenges to overcome. While the diffusion rate of these small molecules to the sensing surface is extremely high, the mass of the analyte is not sufficiently large to produce a measurable response. Other means beyond directly detecting the analyte must, therefore, be used.

The majority of this review focuses on substances the Center for Food Safety and Nutrition (CFSAN), a part of the United States Food and Drug Administration (USFDA), deems as highest risk. While viruses account for the largest percentage of foodborne illness occurrences, little research has been done to detect them. This is largely due to a higher fatality rate associated with bacteria than with viruses, making bacteria an area of greater concern for detection. Typical methods for detecting viruses involve PCR or reverse transcription-PCR techniques [9]. Another area of concern for the food industry is the presence of additives in fruits, vegetables, and meat products. These additives include pesticides or herbicides (e.g., atrazine [10] or simazine [11]), veterinary medicines (e.g., sulfonamides [12–14]), and antibiotics (e.g., hygromycin [15], streptomycin [16], or penicillin [17]). While the presence of these additives is a concern for consumers, the presence of pathogenic bacteria, proteins, and low molecular weight compounds is a higher priority for research and development.

2 Bacteria

Bacteria are, perhaps, the most commonly detected category of foodborne pathogens. Due to their exceptionally large size, and thus extremely low rate of diffusion, detection of live bacteria is extremely problematic. To overcome this limitation, various treatment methods of bacteria are employed. Heat-killing induces morphological changes in the exterior of the bacteria, which could improve detection limits. Alternatively, using ethanol or a detergent to lyse the cells creates smaller fragments that can be detected more readily by decreasing the size of detectable analyte and, thus, increasing the rate of diffusion of the analyte to the sensing surface. Detections of bacteria are separated by species, beginning with the most commonly detected types.

2.1 *Escherichia Coli* O157:H7

E. coli is a bacterium which typically inhabits the intestines of all animals. While the majority of *E. coli* strains are benign, there are several strains of *E. coli* that are capable of causing human illness. *E. coli* O157:H7 is one such strain. *E. coli* O157:H7 produces verotoxins that can cause severe damage to the lining of the intestine. Undercooked or raw ground beef has been implicated in many of the cases of human illness. The infective dose of *E. coli* O157:H7 may be as low as ten cells. Diagnosis of an infection is typically done by isolating toxins from *E. coli* O157:H7 or by isolating the bacterium itself from human stool samples. Confirmation is obtained by isolating the same serotype from the incriminated food sample [6].

Direct detections of *E. coli* have been demonstrated since 1998, when Fratamico et al. demonstrated the detection of viable *E. coli* O157:H7 using a Biacore system [18]. They compared two sensing platforms, both based on attachment via *N*-ethyl-*N*′-(dimethylaminopropyl)carbodiimide hydrochloride (EDC) and *N*-hydroxy-succinimide (NHS) chemistry. Monoclonal or polyclonal antibody was bound directly to the sensing surface or was immobilized on a bound layer of protein A or protein G. Using a sandwich assay, as seen in Fig. 1, a lower limit of detection of $5–7 \times 10^7$ cfu/mL was demonstrated.

Following this first published study on SPR detection of *E. coli* O157:H7, other direct detection studies have been performed. In 2002, Oh et al. used a Multiskop SPR biosensor to detect *E. coli* O157:H7 at concentrations as low as 10^4 cells/mL [19]. This was done using a monoclonal antibody (Mab) immobilized on a protein G surface, similar to the study by Fratamico [18]. In 2003, Oh et al. used the same sensing system as in their previous study, but altered their sensing surface to be an optimized mixed self-assembled monolayer (SAM). This lowered the detection limit of *E. coli* O157:H7 to

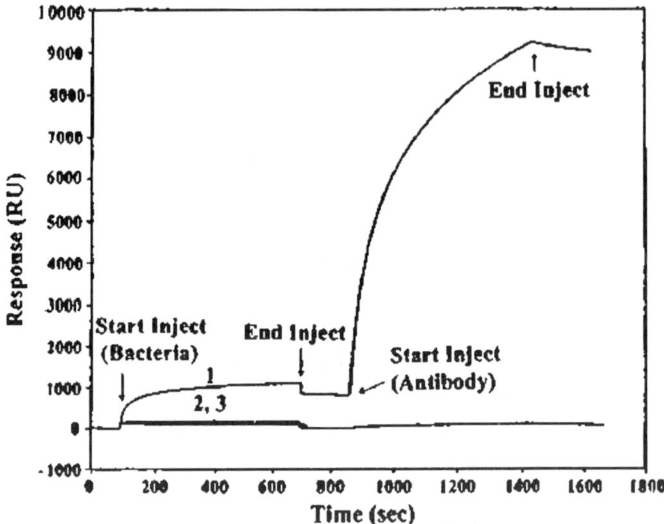

Fig. 1 Overlay plots of sensorgrams showing the interaction of Mab 8-9H (ligand) with *E. coli* O157:H7 (*1*), *S. typhimurium* (*2*) and *Y. entercolitica* (*3*) followed by injection of polyclonal antibody at 50 μg/mL. The bacteria were injected at about 5×10^9 cfu/mL [18]

10^2 cells/mL using the same sensing system [20]. In both studies, no treatment method or amplification protocol was reported. In 2005, Taylor et al. studied various treatment methods and their effect on the detection of *E. coli* O157:H7 with a custom-built sensor [21]. In this study, a Mab was immobilized on the sensing surface via EDC/NHS chemistry, and a sandwich assay with a polyclonal antibody was used for detection. Untreated bacteria were detected at a level comparable to the study by Fratamico, 10^7 cfu/mL. Bacteria treated by heat-killing and subsequent soaking in 70% ethanol showed a detectable limit 10^5 cfu/mL. Lysing bacteria with detergent further improved the detection limit to 10^4 cfu/mL. A summary of detections can be seen in Fig. 2. Taylor et al. also investigated non-specific interactions of bacteria with the sensing surface. Three non-specific, detergent-lysed bacteria, *E. coli* K12, *Salmonella choleraesuis*, and *Listeria monocytogenes*, showed no non-

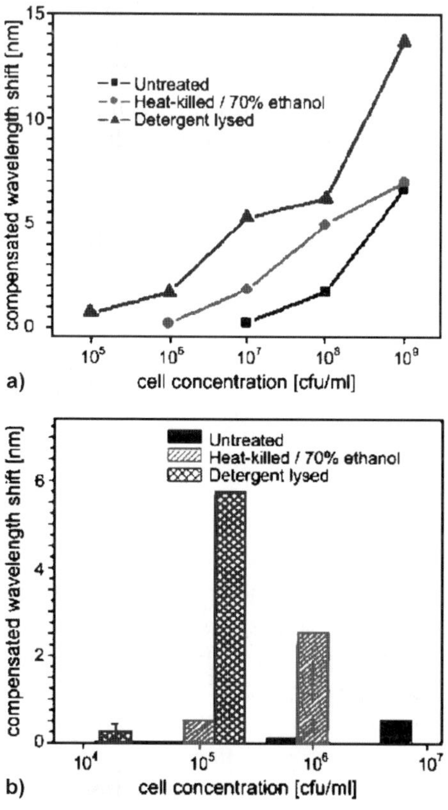

Fig. 2 Resonant wavelength shift versus concentration of bacteria for the detection of *E. coli* O157:H7 comparing untreated, heat-killed then ethanol soaked, and detergent-lysed samples by **a** direct detection and **b** amplification of direct detection by the subsequent exposure to a MAb reactive with *E. coli* O157:H7 (sandwich assay) [21]

specific adsorption to the sensing surface at concentrations of 10^6 cfu/mL, 10^8 cfu/mL, and 10^7 cfu/mL, respectively.

Indirect detections of *E. coli* have also been frequently performed. In 1999, Fratamico et al. used an inhibition assay to detect *E. coli* O157:H7 [22]. Polyclonal antibodies to *E. coli* were incubated with viable *E. coli*. Bacteria were centrifuged out, and the supernatant containing unbound antibody was flowed across the sensing surface. An anti-Fab antibody immobilized to the sensor surface was used to capture the free antibody. Limits of detection using the inhibition assay were between 10^6 cfu/mL and 10^7 cfu/mL. Also in 1999, Kai et al. demonstrated the use of an SPR sensor to detect PCR products of the *Escherichia coli* O157:H7 genome [23]. A biotinylated probe DNA sequence was immobilized on a streptavidin-coated surface. To this, a double stranded target DNA was bound. The target DNA had a probe site located in the 3′-terminus that was single stranded. Detection limits of bacteria were not reported. In 2000, Kai et al. demonstrated detection of *E. coli* O157:H7 in stool samples using PCR [24]. A biotinylated PNA was immobilized on a streptavidin-coated surface and used as the probe strand. *E. coli* O157:H7 was detectable at 10^2 cfu/0.1 g of stool sample. Spangler et al. detected *E. coli* heat-labile enterotoxin in 2001 [25]. In this study, ganglioside GM_1 was immobilized directly on a gold surface. Direct detections of the enterotoxin were demonstrated for concentrations from 70 nM to 600 nM.

2.2 *Salmonella* spp.

Salmonella spp. is an infectious bacterium typically found in raw meats, poultry, and seafood, as well as eggs, milk, and dairy products. The infective dose of *Salmonella* spp. can be as few as 15–20 cells, depending on the size and health of the individual. Diagnosis of an infection is typically done by isolating the bacteria from stool samples [6].

In 2001, Koubová et al. demonstrated detection of *Salmonella enteritidis* using a custom-built SPR system [26]. A double layer of antibodies was physisorbed onto a bare gold surface and crosslinked with gluteraldehyde. Using this sensing surface, direct detections of heat-killed, ethanol soaked *S. enteritidis* as low as 10^6 cells/mL were demonstrated. Bokken et al. detected *Salmonella* strains from group A, B, D, and E, according to Kauffmann–White typing, using a Biacore system in 2003 [27]. Antibodies were immobilized on the dextran chip surface using EDC/NHS chemistry. Detections of heat-killed *Salmonella* serotypes were run in mixtures of non-*Salmonella* bacteria using a sandwich assay. *Salmonella* at concentrations of 1.7×10^5 cfu/mL and higher were detectable in mixtures containing 10^8 cfu/mL non-*Salmonella* bacteria. Oh et al. showed detections of *Salmonella typhimurium* in 2004 [28] using a surface based on the same protein G construct used in their work with *E. coli* O157:H7 [19]. Mabs to *S. typhimurium* were immobilized on the protein G surface. Detections from 10^2 cfu/mL up to 10^9 cfu/mL were shown. Detections of *Salmonella paratyphi* were also demonstrated by Oh et al. in 2004 [29]. A self-assembled monolayer of a thiol-substituted protein G was made. Mab was then adsorbed on the surface and used for detection. Detection of *S. paratyphi* was shown down to concentrations of 10^2 cfu/mL. In both studies by Oh et al. no amplification method was discussed, and no discussion of treatment methods for the bacteria was given.

2.3
Listeria Monocytogenes

Listeria monocytogenes has been isolated from raw fish, cooked crab, raw and cooked shrimp, raw lobster, surimi, and smoked fish. While the infective dose of *L. monocytogenes* is unknown, it is believed to be as low as fewer than 1000 organisms. Current techniques for the diagnosis of listeriosis involve culturing the organism from blood or cerebrospinal fluid [6]. In conjunction with their study of *S. enteritidis* in 2001, Koubová et al. also demonstrated detection of *L. monocytogenes* [26]. With the same sensing surface chemistry, a double layer of physisorbed antibodies, direct detections were demonstrated from 10^7 cells/mL to 10^9 cells/mL. The *Listeria* bacteria were heat killed for this study.

In 2004, Leonard et al. used an inhibition assay to detect the presence of whole cell *L. monocytogenes* in solution [30]. A commercial goat anti-rabbit polyclonal antibody was immobilized using EDC/NHS chemistry on a dextran surface. Solutions of known concentrations of *L. monocytogenes* were incubated with rabbit anti-*Listeria* antibodies. Cells and bound antibodies were then centrifuged out of solution and the unbound antibodies remaining in solution were detected by the sensing surface. Using this method, detections of 10^5 cells/mL were demonstrated.

2.4
Campylobacter Jejuni

Campylobacter jejuni is thought to be the leading cause of bacterial diarrheal illness in the United States. While not carried by healthy individuals in the United States or Europe, *C. jejuni* has been isolated from healthy cattle and chickens, as well as non-chlorinated water sources. The infective dose is considered small, with studies suggesting that as few as 400–500 bacteria may cause illness. Typical diagnosis of campylobacteriosis is done by isolation from human stool samples [6].

Taylor et al. recently demonstrated detections of *C. jejuni* [31]. These detections were done in conjunction with a study demonstrating simultaneous detection of multiple bacteria, described later in this review. Detection was done using a sandwich assay and antibodies immobilized via biotin–

streptavidin interactions. Detections for heat-killed *C. jejuni* were shown for buffer solutions containing only *C. jejuni*, as well as a mixture of *C. jejuni* and other non-specific bacteria. Limits of detection in both cases were established at concentrations of 5×10^4 cfu/mL. Detections were also performed in the applicable real-world complex matrix, apple juice. Detections at the native pH of apple juice, pH 3.7, as well as detections in apple juice at physiological pH, 7.4, were performed. Detection limits for both cases were comparable to those demonstrated in buffer.

2.5 *Clostridium Perfringens*

Clostridium perfringens is found in soil, sediments, and areas subject to human or animal fecal pollution. Perfringens food poisoning is caused by ingestion of large numbers of toxin-producing *C. perfringens* bacteria. The infective dose is greater than 10^8 vegetative cells. Diagnosis of perfringens poisoning is confirmed by detecting the toxin in the feces of patients [6].

Hsieh et al. detected the β-toxin produced by *C. perfringens* in 1998 [32]. Mabs were immobilized on a dextran surface using EDC/NHS chemistry. Detections showed a working range from 1/1.02 to 1/220 dilutions. While amplification was tested with a sandwich assay, it does not appear it was used in production of the detection curve.

2.6 *Yersinia Enterocolitica*

Yersinia enterocolitica has been isolated in ponds, lakes, meats, ice cream, and milk. Typical symptoms of yersiniosis include gastroenteritis, diarrhea, fever, and abdominal pain. While the infective dose is unknown, diagnosis of yersiniosis is done by isolating the organism from the human host's feces, blood, or vomit [6].

In 2005, Oh et al. performed detections of *Y. Enterocolitica* [33]. A Mab immobilized on a protein G surface similar to their study of *E. coli* O157:H7 in 2001 and *S. typhimurium* in 2003 was used. Detections from 10^2 cfu/mL to 10^7 cfu/mL were observed. As with previous studies from this group, no amplification or treatment of the bacteria was reported.

2.7 Multiple Bacteria Detections

Most of the work that has been done with bacteria over the past several years has focused on improving the detection limits of the SPR sensors. This is an important focus of research, as most of the infective doses of foodborne bacteria are still below or very near the current lower detection thresholds of the most sensitive instruments. Another focus of the sensor technology has been to expand from detection of only one type of bacteria to the detection of multiple bacteria. Few studies have been done showing a multichannel capacity for foodborne pathogens.

In 2005 Oh et al. published a study showing detections of individual bacteria on a multichannel SPR sensor [34]. Using a protein G-immobilized surface, Mabs to four individual bacteria (*E. coli* O157:H7, *S. typhimurium*, *Listeria pheumophila*, *Y. enterocolitica*) were immobilized into individual sensing channels. Each of the four bacteria was then flowed in succession across each of the sensing channels. Sensor responses were only performed for 10^5 cfu/mL of each of the bacteria. Responses showed good specificity within the sensing channel, as seen in Fig. 3. No amplification or treatment of the bacteria was reported.

Taylor et al. demonstrated the simultaneous detection of four different bacteria, *E. coli* O157:H7, *C. jejuni*, *S. typhimurium* and *L. monocytogenes*, in buffer and apple juice on a custom-built SPR system [31]. All bacteria were heat-killed and ultrasonicated prior to detection. Detections were performed on individual bacteria, as well as in mixtures of all four bacteria. Simultaneous detection of individual bacteria in the mixtures showed good agreement with detections of individual bacteria in buffer. Detections of individual bacteria and mixtures were also performed in apple juice at both

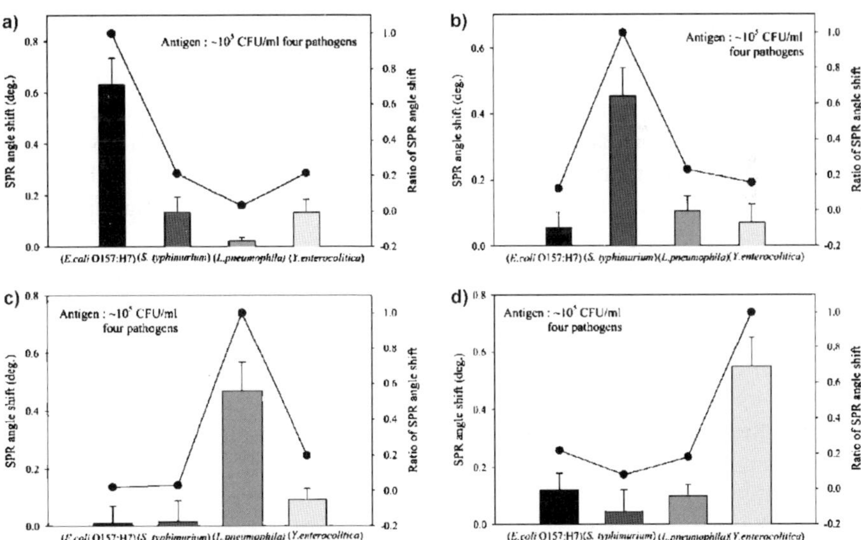

Fig. 3 The response of four individually functionalized spots of Mab against four pathogens flowed in succession at concentrations of 10^5 cfu/mL. **a** Mab against *E. coli* O157:H7; **b** Mab against *S. typhimurium* spot; **c** Mab against *L. pheumophila* spot; and **d** Mab against *Y. entercolitica* spot. The *bar graph* indicates the SPR angle shift, while the *circles* represent the ratios of SPR angle shift

a native pH of 3.7, as well as apple juice adjusted to a physiological pH of 7.4. Adjusting the pH altered the sensor response, but did not affect the attainable lower detection limits. Limits of detection for all four cases were established at 10^4 cfu/mL, 5×10^4 cfu/mL, 5×10^4 cfu/mL, and 10^4 cfu/mL for *E. coli* O157:H7, *C. jejuni*, *S. typhimurium*, and *L. monocytogenes*, respectively.

3
Proteins

Toxic foodborne proteins are typically secreted from infectious bacteria. These proteins are typically toxic in extremely low doses. Because of the molecular weight of the proteins (ranging from 5 kDa to 150 kDa), diffusion to the surface is not problematic. Their mass allows for sensor responses from direct binding at higher concentrations. For lower analyte concentrations, a sandwich assay is typically used to improve the lower detection limits for various analytes.

3.1
Staphylococcal Enterotoxins

Staphylococcal enterotoxins (SEs) are the most widely studied of the toxic foodborne proteins. Although these toxins are produced by various strains of *Staphylococcus*, evidence has shown they are primarily produced by the *Staphylococcus aureus* strain. Currently there are nine enterotoxins (A, B, C, D, E, G, H, I, J) that have been identified in a wide variety of food products: meat, poultry and egg products, milk and dairy products, as well as bakery products [35]. The infective dose of toxins is estimated to be 0.1 μg/kg body mass [36]. Detection of the presence of SEs is typically done through isolation in the suspected food source [6].

Nedelkov et al. demonstrated the detection of staphylococcal enterotoxin B (SEB) in 2000 [37]. Antibodies to SEB were immobilized on the sensing surface using EDC/NHS chemistry. This study demonstrated direct detection of concentrations ranging from 1 ng/mL to 100 ng/mL. Detections from 1 ng/mL to 100 ng/mL were also demonstrated in a milk matrix, as well as in a mushroom extract solution. In 2002, Homola et al. demonstrated improved detection limits of SEB [38] (see Fig. 4). Using antibody immobilized via EDC/NHS chemistry, detections ranging from 0.5 ng/mL to 50 ng/mL were demonstrated. Detections were performed directly and using amplification with a secondary antibody. Detections for concentrations as low as 0.5 ng/mL were also demonstrated in milk using amplification.

In 2002, Naimushin et al. detected SEB using a custom-built SPR sensor [39]. Instead of using EDC/NHS chemistry between antibody and a SAM, EDC/NHS chemistry was used to link the antibody to a gold-binding pep-

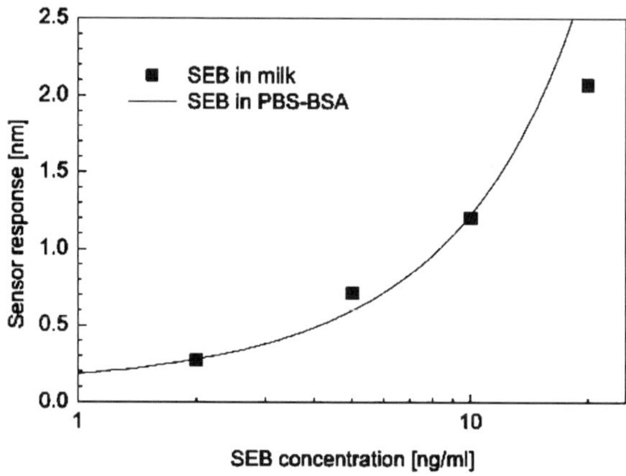

Fig. 4 Equilibrium surface plasmon resonance sensor response to staphylococcal enterotoxin B (SEB) in a solution of BSA in phosphate buffered saline (BSA-PBS). Reference-compensated equilibrium sensor response to different concentrations of SEB in BSA-PBS solution for direct and sandwich detection modes (a-SEB concentration 3 μg/mL in BSA-PBS) [38]

tide immobilized on the gold surface. Direct detection of SEB was observed for concentrations ranging from 0.2 nM (∼ 5.7 ng/mL) to 75 nM (2.1 μg/mL) in buffer and as low as 0.5 nM (11.4 ng/mL) and 1 nM (22.8 ng/mL) in urine and seawater, respectively. Using a one-step amplification, concentrations of 50 pM (1.4 ng/mL) and 20 pM (0.6 ng/mL) were detected in seawater and buffer, respectively. Using two amplification steps, concentrations of 100 fM (2.8 pg/mL) were demonstrated in buffer. Also in 2002, Slavík et al. performed a study using a fiber optic sensor for detection of SEB [40]. A double layer of antibodies was physisorbed onto a tantalum pentoxide surface and cross-linked with gluteraldehyde. Direct detections with this sensor were demonstrated for concentrations between 10 ng/mL and 100 ng/mL. No amplification was used.

In 2003, Nedelkov and Nelson demonstrated the ability to recycle their sensing surface, making multiple detections on the same chip possible [41]. Detections at 1 ng/mL done with three recycle steps between the detections showed a good correlation. Detections in mushroom extract were performed and showed similar limits to the previous study in 2000. Medina also detected SEB using a sandwich assay in 2003 [42]. SEB antibody was immobilized on the sensing surface. Detection limits of 2.5 ng/mL were demonstrated in both buffer and ham tissue extract. Naimushin et al. studied the effect of temperature on the detection of SEB in lake water in 2003 [43]. Using the same sensing platform as their 2002 study, direct detections of 2 nM (45.6 ng/mL) were demonstrated. A temperature effect on the initial binding rate of SEB during detections was also shown. A maximum rate was seen at ∼ 45 °C for their system.

Medina used an inhibition assay to detect SEB in 2005 [44]. SEB was immobilized on the sensing surface. Known concentrations of SEB were then incubated with anti-SEB antibodies, and the incubation solution was passed across the sensing surface. Detections in buffer were seen for concentrations ranging from 0.78 ng/mL to 50 ng/mL. In whole and skim milk, detections were demonstrated for 0.312 ng/mL to 25 ng/mL.

3.2 Botulinum Neurotoxins

Botulinum neurotoxins (BoNTs) are perhaps the most lethal toxins known. BoNTs are a set of seven serotypes (A, B, C, D, E, F and G) that are produced almost exclusively by the bacteria *Clostridium botulinum*. Serotypes C and D are found in birds and non-human mammals. Types A, B, E and F have been implicated in human cases of botulism. LD50 values for BoNTs range from 1.1 to 2.5 ng/kg body weight [7]. BoNTs have been associated with a variety of foods, including honey, chili, and hash browns. Isolation of toxins from the suspected food is the current means of diagnosis.

Ladd et al. have recently demonstrated the simultaneous detection of three serotypes of BoNT using a custom-built SPR sensor (submitted for publication). Detections of serotypes A, B, and F were done using a sandwich assay with polyclonal antibodies immobilized via biotin–streptavidin interactions. The lowest concentrations detected for serotypes A, B, and F in buffer were 1 ng/mL, 1 ng/mL, and 0.5 ng/mL, respectively. Detections performed in a 20% honey solution showed good agreement with detections performed in buffer.

4 Low Molecular Weight Compounds

Small molecules pose a far different set of challenges for an SPR sensor than bacteria. While the diffusion rate of the small analytes is quite large, their low molecular weight does not cause a significant increase in the local refractive index near the sensing surface. Because of this problem, various strategies have been developed, the most common of which is the use of an inhibition or competitive assay.

4.1 Domoic Acid

Domoic acid (DA) is a neuroexcitatory toxin typically produced by planktonic algae. Shellfish, especially mussels, become contaminated with DA upon eating the toxin-containing algae. Ingestion of contaminated shellfish by humans

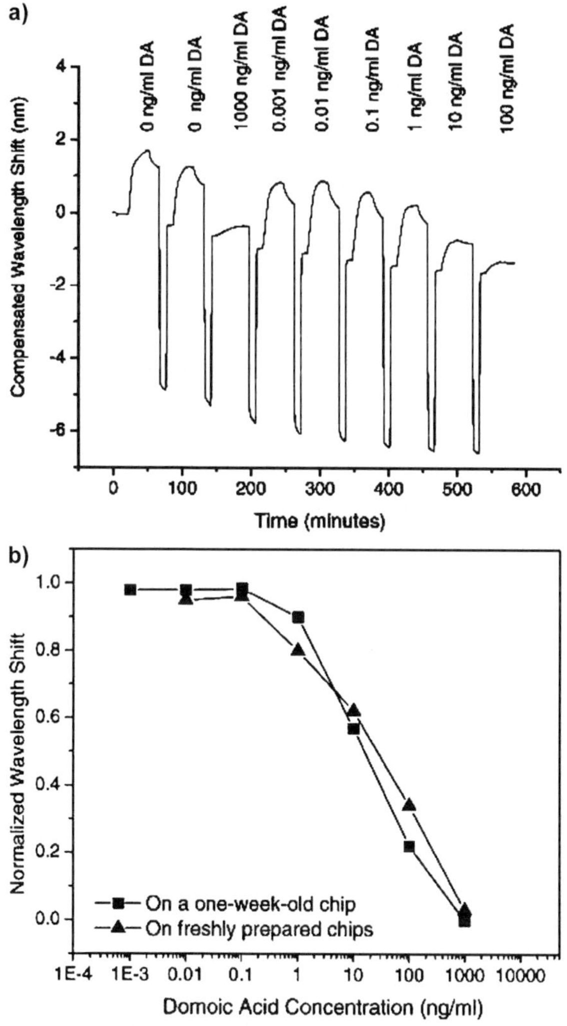

Fig. 5 **a** SPR sensorgrams of different concentrations of DA incubated with 1.25 mg/mL anti-DA antibodies in solution measured on a chip, which was stored for 1 week and regenerated using a 50 mM NaOH solution after each measurement. **b** Comparison of the detection curves measured on freshly prepared chips and on the stored and regenerated chip [46]

can cause an intoxication syndrome known as amnesic shellfish poisoning (ASP). Symptoms of ASP include vomiting and cramps, as well as seizures and temporary or permanent memory loss [6].

In 2004, Lotierzo et al. demonstrated the use of a Biacore 3000 system for the detection of domoic acid [45]. Their sensor platform consisted of a molecularly imprinted polymer film synthesized by direct photografting. Using a competitive assay, detections from 2 ng/mL to 3.3 μg/mL were demonstrated. The surface was regenerated after each detection and showed good stability through 30 regenerations. Yu et al. also demonstrated the detection of domoic acid using a custom-built SPR sensor in 2005. DA was immobilized on a mixed SAM of oligo (ethylene glycol) (OEG)-containing alkanethiols using EDC/NHS chemistry [46] (see Fig. 5). The mixed SAM was tailored to improve the nonfouling properties of the sensor chip. Using an inhibition assay, DA was detected in a range from 0.1 ng/mL to 1000 ng/mL. Detection responses following regeneration of the chip surface showed good agreement with detections done on a freshly mounted chip surface.

4.2 Mycotoxins

Mycotoxins are low molecular weight metabolites produced by mold genera. Production of these toxins is commonly associated with the bacterial species *Aspergillus*, *Penicillium*, and *Fusarium*. Mycotoxins cause adverse health effects ranging from nausea and vertigo to carcinogenic and genotoxic effects [6]. Three of the most commonly occurring mycotoxins are fumonisins, deoxynivalenol, and aflatoxins.

4.2.1 Fumonisins

Fumonisin B_1 (FB_1) is the most prevalent fumonisin in contaminated food and is believed to be the most toxic to afflicted animals. Though a variety of adverse effects have been observed in livestock, there is currently no direct evidence of the effects of fumonisins on humans. Some inconclusive observations show that fumonisins could be carcinogenic. Fumonisins are associated mainly with corn products [6].

In 1998, Mullett et al. used a custom-built SPR biosensor to detect FB_1 [47]. Antibody was physisorbed to the bare gold sensing surface. FB_1 was detected directly at concentrations ranging from 50 ng/mL to 100 μg/mL.

4.2.2 Deoxynivalenol

Deoxynivalenol is found most commonly on grains such as rye, rice, wheat, and corn. A proposed tolerable daily intake was defined at 0.5 μg/kg body weight, which corresponds to concentrations of 100–500 μg/kg in wheat.

Tüdös et al. demonstrated the use of a Biacore-Q system for the detection of deoxynivalenol in wheat samples using an inhibition assay [48]. Deoxynivalenol was conjugated to casein and immobilized on a CM5 chip using

EDC/NHS chemistry. Detections at concentrations ranging from 0.05 ng/mL to 1000 ng/mL were demonstrated in buffer. Natural contaminations in wet and dry samples of wheat were also tested and found to produce results similar to those obtained using other techniques. The effect of antibody-sample incubation time was also studied. Choi et al. demonstrated the use of single-chain variable fragment antibodies for the detection of deoxynivalenol in 2004 [49]. A deoxynivalenol–hemiglutarate–horseradish peroxidase conjugate was immobilized on a CM5 chip using amine coupling. scFvs were then flowed across the sensor surface. While no detection limits were established, affinity constants were obtained.

4.2.3 Aflatoxins

Aflatoxins are a group of highly toxic and carcinogenic compounds produced by certain strains of *Aspergillus flavus* and *Aspergillus parasiticus*. Contamination occurs most frequently in tree nuts, peanuts, and other oilseeds, including corn. The most toxic and most predominant of these toxins is aflatoxin B_1. LD50 values for aflatoxins range from 0.5 to 10 mg/kg body weight. Diagnosis is typically done by isolating the toxin in the suspected food.

Daly et al. detected aflatoxin B_1 using an inhibition assay on a Biacore system in 2000 [50]. Aflatoxin B_1 was conjugated to BSA and immobilized on a dextran surface using EDC/NHS chemistry. Aflatoxin-containing samples were then incubated with antibody in solution. This solution was then passed across the sensor surface. Detections were demonstrated with this inhibition assay for concentrations ranging from 3 ng/mL to 100 ng/mL. Dunne et al. demonstrated in 2005 the ability to detect aflatoxin B_1 using single-chain antibody fragments (scFvs) [51]. An aflatoxin B_1 derivative was immobilized on a CM5 chip. Aflatoxin B_1 was incubated with monomeric or dimeric scFvs, and the solution was passed over the sensor surface. Detections ranging from 375 pg/mL to 12 ng/mL were demonstrated for the monomeric scFv. The dimeric scFvs were able to detect concentrations ranging from 190 pg/mL to 24 ng/mL. The sensor surface also showed good agreement between 75 regeneration cycles.

5 Summary and Future Trends

SPR biosensors have been used to detect a wide range of food-related analytes. Table 1 provides a summary of the studies listed in this review. SPR sensors are at the forefront of sensing technologies capable of real-time, quantitative detections. This makes them quite suitable as on-line monitoring systems in food processing plants. Trends in the technology are pushing

Table 1 Overview of SPR sensor detections of foodborne pathogens

Specific analyte designation	Sensor system	Comments	Detection matrix	Sensitivity	Assay format	Reference
***Escherichia coli* O157:H7**						
	Biacore	Live	HBS, pH 7.4	$5–7 \times 10^7$ cfu/mL	Sandwich	[18]
	Multiskop			10^4 cells/mL		[19]
	Multiskop			10^2 cells/mL		[20]
		Heat-killed	PBS, pH 7.4	10^6 cfu/mL	Sandwich	
	Custom-built	Heat-killed & ethanol-soaked	PBS, pH 7.4	10^5 cfu/mL	Sandwich	[21]
		Detergent-lysed	PBS, pH 7.4	10^4 cfu/mL	Sandwich	
	Multiskop			10^5 cfu/mL		[34]
	Biacore	Live	HBS, pH 5.0	$10^6–10^7$ cfu/mL	Inhibition	[22]
	Biacore 2000	PCR				[23]
	Biacore 2000	PCR	Stool sample	10^2 cfu/0.1 g		[24]
	Spreeta	Enterotoxin detection	PBS, pH 7.2	6 μg/mL		[25]
			PBS, pH 7.4	10^4 cfu/mL	Sandwich	
	Custom-built	Heat-killed & ultrasonicated	Apple juice, pH 3.7	10^4 cfu/mL	Sandwich	[31]
			Apple juice, pH 7.4	10^4 cfu/mL	Sandwich	
***Salmonella* spp.**						
S. enteritidis	Custom-built	Heat-killed, ethanol-soaked	PBS, pH 7.4	10^6 cells/mL		[26]
	Biacore	Heat-killed	Mixture of non-*Salmonella* bacteria	1.7×10^5 cfu/mL	Sandwich	[27]
S. paratyphi	Multiskop			10^2 cfu/mL	Direct	[29]
S. typhimurium	Multiskop			10^2 cfu/mL	Direct	[28]
S. typhimurium	Multiskop			10^5 cfu/mL	Direct	[34]

Table 1 (continued)

Specific analyte designation	Sensor system	Comments	Detection matrix	Sensitivity	Assay format	Reference
			PBS, pH 7.4	5×10^4 cfu/mL	Sandwich	
S. typhimurium	Custom-built	Heat-killed & ultrasonicated	Apple juice, pH 3.7	5×10^4 cfu/mL	Sandwich	[31]
			Apple juice, pH 7.4	5×10^4 cfu/mL	Sandwich	
Listeria monocytogenes						
	Custom-built	Heat-killed	PBS, pH 7.4	10^6 cells/mL	Direct	[26]
	Biacore 3000	Heat-killed		10^5 cells/mL	Inhibition	[30]
	Multiskop			10^5 cfu/mL	Direct	[34]
			PBS, pH 7.4	10^4 cfu/mL	Sandwich	
	Custom-built	Heat-killed & ultrasonicated	Apple juice, pH 3.7	10^4 cfu/mL	Sandwich	[31]
			Apple juice, pH 7.4	10^4 cfu/mL	Sandwich	
Campylobacter jejuni						
			PBS, pH 7.4	5×10^4 cfu/mL	Sandwich	
	Custom-built	Heat-killed & ultrasonicated	Apple juice, pH 3.7	5×10^4 cfu/mL	Sandwich	[31]
			Apple juice, pH 7.4	5×10^4 cfu/mL	Sandwich	
Clostridium perfringens						
	Biacore			1/220 dilution	Direct	[32]
				β-Toxin		
Yersinia enterocolitica						
	Multiskop			10^2 cfu/mL	Direct	[33]
	Multiskop			10^5 cfu/mL	Direct	[34]

Table 1 (continued)

Specific analyte designation	Sensor system	Comments	Detection matrix	Sensitivity	Assay format	Reference
Staphylococcal enterotoxins (SEs)						
			HBS-EP	1 ng/mL	Direct	
SEB	Biacore		Milk	1 ng/mL	Direct	[37]
			Mushroom extract	1 ng/mL	Direct	
SEB	Custom-built		PBS, pH 7.4	0.5 ng/mL	Sandwich	[38]
			Milk	0.5 ng/mL	Sandwich	
SEB	Custom-built		TTBS, pH 8.0	2.8 ng/mL	Double Amplification Sandwich	[39]
			Milk	N/A	N/A	
			Urine	11.4 ng/mL	Sandwich	
			Sea water	1.4 ng/mL	Sandwich	
SEB	Custom fiber optic		PBS, pH 7.4	10 ng/mL	Direct	[40]
SEB	Biacore		HBS-EP	1 ng/mL	Direct	[41]
			Mushroom extract	1 ng/mL	Direct	
SEB			Buffer	2.5 ng/mL	Sandwich	[42]
			Ham tissue extract	2.5 ng/mL	Sandwich	
SEB	Custom-built		Lake water	45.6 ng/mL	Sandwich	[43]
SEB			Buffer	0.78 ng/mL	Inhibition	[44]
			Milk	0.312 ng/mL	Inhibition	
Botulinum neurotoxins (BoNTs)						
BoNTs A, B, F	Custom-built		PBS, pH 7.4		Sandwich	
			20% Honey		Sandwich	

Table 1 (continued)

Specific analyte designation	Sensor system	Comments	Detection matrix	Sensitivity	Assay format	Reference
Domoic acid						
	Biacore 3000	MIP film	HBS-EP	2 ng/mL	Competitive	[45]
	Custom-built		PBS, pH 7.4	0.1 ng/mL	Inhibition	[46]
Mycotoxins						
Fumonisins	Custom-built		PBS, pH 7.4	50 ng/mL	Direct	[47]
Deoxynivalenol	Biacore-Q™		HBS, pH 7.4 Ground wheat	0.05 ng/mL	Inhibition Inhibition	[48]
Deoxynivalenol	Biacore 3000	scFV			Direct	[49]
Aflatoxin B_1	Biacore 1000		HBS, pH 7.4	3 ng/mL	Inhibition	[50]
Aflatoxin B_1	Biacore™ 3000	Monomeric scFV Dimeric scFV	PBS, 5% methanol PBS, 5% methanol	375 pg/mL 190 pg/mL	Inhibition Inhibition	[51]

towards further automation of systems, miniaturization of sensing systems, and high-throughput capabilities. Automated systems would allow for deployment of a sensor for continuous monitoring of food-processing plants, or for environmental monitoring. A portable SPR sensing system enabling rapid detections of foodborne pathogens would greatly benefit producers, processors, and distributors in the food industry. High throughput systems would allow screening for a large number of analytes within a single sample. Development of these SPR systems will require significant advances in miniaturization of the sensing platforms, integration with microfluidic devices, and the development of robust, highly specific biomolecular recognition elements.

Acknowledgements This project was funded by grants from the United States Food and Drug Administration (FD-U-002250) and the United States Department of Agriculture (CSREES 2005-01916).

References

1. Mead PS, Slutsker L, Dietz V, McCaig LF, Bresee JS, Shapiro C, Griffin PM, Tauxe RV (1999) Emerg Infect Dis 5:607
2. Gracias KS, McKillip JL (2004) Can J Microbiol 50:883
3. Nakano S, Kobayashi T, Funabiki K, Matsumura A, Nagao Y, Yamada T (2004) J Food Protect 67:1271
4. Gunasekera TS, Veal DA, Attfield PV (2003) Int J Food Microbiol 85:269
5. Croci L, Delibato E, Volpe G, Palleschi G (2001) Anal Lett 34:2597
6. Center for Food Safety and Nutrition & United States Food and Drug Administration (1999) Foodborne pathogenic microorganisms and natural toxins handbook. US FDA
7. Jin A, Weekly S (2003) Botulinum toxin fact sheet. Lawrence Livermore National Laboratory
8. United States Department of Homeland Security (2004) Homeland Security Advanced Research Projects Agency (HSARPA) Broad Agency Announcement (BAA) 05–06
9. Carter MJ (2005) J Appl Microbiol 98:1354
10. Minunni M, Mascini M (1993) Anal Lett 26:1441
11. Harris RD, Luff BJ, Wilkinson JS, Piehler J, Brecht A, Gauglitz G, Abuknesha RA (1999) Biosens Bioelectron 14:377
12. Haasnoot W, Bienenmann-Ploum M, Kohen F (2003) Anal Chim Acta 483:171
13. Crooks SRH, Baxter GA, O'Connor MC, Elliot CT (1998) Analyst 123:2755
14. Sternesjo A, Mellgren C, Bjorck L (1995) Anal Biochem 226:175
15. Medina MB (1997) J Agric Food Chem 45:389
16. Baxter GA, Ferguson JP, O'Connor MC, Elliott CT (2001) J Agric Food Chem 49:3204
17. Gustavsson E, Bjurling P, Sternesjo A (2002) Anal Chim Acta 468:153
18. Fratamico PM, Strobaugh TP, Medina MB, Gehring AG (1998) Biotechnol Tech 12:571
19. Oh BK, Kim YK, Bae YM, Lee WH, Choi JW (2002) J Microbiol Biotechnol 12:780
20. Oh BK, Lee W, Lee WH, Choi JW (2003) Biotechnol Bioprocess Eng 8:227
21. Taylor AD, Yu QM, Chen SF, Homola J, Jiang SY (2005) Sensor Actuator B Chem 107:202
22. Fratamico PM, Strobaugh TP, Medina MB, Gehring AG (1999) A surface plasmon resonance biosensor for real-time immunologic detection of *Escherichia coli* O157:H7.

In: Tunick M, Fratamico PM, Palumbo SA (eds) New techniques in the analysis of foods. Kluwer Academic, New York, p 103
23. Kai E, Sawata S, Ikebukuro K, Iida T, Honda T, Karube I (1999) Anal Chem 71:796
24. Kai E, Ikebukuro K, Hoshina S, Watanabe H, Karube I (2000) FEMS Immunol Med Microbiol 29:283
25. Spangler BD, Wilkinson EA, Murphy JT, Tyler BJ (2001) Anal Chim Acta 444:149
26. Koubova V, Brynda E, Karasova L, Skvor J, Homola J, Dostalek J, Tobiska P, Rosicky J (2001) Sensor Actuator B Chem 74:100
27. Bokken G, Corbee RJ, van Knapen F, Bergwerff AA (2003) FEMS Microbiol Lett 222:75
28. Oh BK, Kim YK, Park KW, Lee WH, Choi JW (2004) Biosens Bioelectron 19:1497
29. Oh BK, Lee W, Kim YK, Lee WH, Choi JW (2004) J Biotechnol 111:1
30. Leonard P, Hearty S, Quinn J, O'Kennedy R (2004) Biosens Bioelectron 19:1331
31. Taylor AD, Ladd J, Yu Q, Chen S, Homola J, Jiang S (2006) Biosens Bioelectron (in press)
32. Hsieh HV, Stewart B, Hauer P, Haaland P, Campbell R (1998) Vaccine 16:997
33. Oh BK, Lee W, Chun BS, Bae YM, Lee WH, Choi JW (2005) Colloids Surf A Physicochem Eng Asp 257–58:369
34. Oh BK, Lee W, Chun BS, Bae YM, Lee WH, Choi JW (2005) Biosens Bioelectron 20:1847
35. Le Loir Y, Baron F, Gautier M (2003) Genet Mol Res 2:63
36. Evenson ML, Hinds MW, Bernstein RS, Bergdoll MS (1988) Int J Food Microbiol 7:311
37. Nedelkov D, Rasooly A, Nelson RW (2000) Int J Food Microbiol 60:1
38. Homola J, Dostalek J, Chen SF, Rasooly A, Jiang SY, Yee SS (2002) Int J Food Microbiol 75:61
39. Naimushin AN, Soelberg SD, Nguyen DK, Dunlap L, Bartholomew D, Elkind J, Melendez J, Furlong CE (2002) Biosens Bioelectron 17:573
40. Slavik R, Homola J, Brynda E (2002) Biosens Bioelectron 17:591
41. Nedelkov D, Nelson RW (2003) Appl Environ Microbiol 69:5212
42. Medina MB (2003) J Rapid Methods Automat Microbiol 11:225
43. Naimushin AN, Soelberg SD, Bartholomew DU, Elkind JL, Furlong CE (2003) Sensor Actuator B Chem 96:253
44. Medina MB (2005) J Rapid Methods Automat Microbiol 13:3
45. Lotierzo M, Henry OYF, Piletsky S, Tothill I, Cullen D, Kania M, Hock B, Turner APF (2004) Biosens Bioelectron 20:145
46. Yu QM, Chen SF, Taylor AD, Homola J, Hock B, Jiang SY (2005) Sensor Actuator B Chem 107:193
47. Mullett W, Lai EPC, Yeung JM (1998) Anal Biochem 258:161
48. Tudos AJ, Lucas-van den Bos ER, Stigter ECA (2003) J Agric Food Chem 51:5843
49. Choi GH, Lee DH, Min WK, Cho YJ, Kweon DH, Son DH, Park K, Seo JH (2004) Protein Expr Purif 35:84
50. Daly SJ, Keating GJ, Dillon PP, Manning BM, O'Kennedy R, Lee HA, Morgan MRA (2000) J Agric Food Chem 48:5097
51. Dunne L, Daly S, Baxter A, Haughey S, O'Kennedy R (2005) Spectrosc Lett 38:229

Springer Ser Chem Sens Biosens (2006) 4: 229–247
DOI 10.1007/5346_022

Published online: 8 July 2006

SPR Biosensors for Medical Diagnostics

Hana Vaisocherová · Jiří Homola (✉)

Institute of Radio Engineering and Electronics, Prague, Czech Republic
homola@ure.cas.cz

Keywords Biosensor · Cancer markers · Detection of antibodies · Detection of disease biomarkers · Detection of hormones · Disease diagnostics · Heart attack markers · Monitoring drug levels · Sensor · Surface plasmon resonance

1 Introduction

Advances in the life sciences (e.g., genomics, proteomics, molecular engineering) have improved the treatment of a wide range of diseases, resulting in an improved public health and a longer life expectancy. In developed countries, lifestyle diseases such as cardiovascular disease have become a major public health concern and one of the leading causes of mortality. Modern health care increasingly involves diagnostic methods based on the monitoring of disease biomarkers in bodily fluids, as some of these markers allow for identification of a disease at its very early stage even before its symptoms can be found. In addition, the monitoring of concentrations of biological markers in bodily fluids can help determine predispositions for the disease and disease progression. Furthermore, the detection of biological markers can help redefine the diseases and their therapies by shifting the emphasis of traditional practices of depending on symptoms and morphology to a more rational objective molecular basis. While biomarkers for certain diseases are established and already in clinical use (e.g., prostate-specific antigen for prostate cancer), the search for reliable diagnostic biomarkers for other diseases continues [1].

Currently, most methods for the determination of biomarkers in bodily fluids are carried out in hospitals or specialized laboratories. These include enzyme-linked immunosorbent assay (ELISA), chemiluminescent, immunofluorescent, radiological, and microscopic assays. The immunoassays utilize antibodies as biomolecular recognition elements and, due to the advances in antibody engineering and synthesis of humanized antibodies [2], present one of the fastest growing diagnostic technologies. However, these methods are rather laborious and time-consuming and offer limited automation and integration of the various operational steps [3]. Detection formats typically require labeling and the use of additional reagents to report binding of the analyte to the receptor [4]. The labeling prolongs assay time, increases costs, and can disturb receptor binding sites leading to false negatives. Moreover, fluorescent compounds are invariably hydrophobic, and in many screening methods, background is a significant problem potentially leading to false positives. Furthermore, at present there are no accepted immunoassay tests for certain serious diseases (e.g., cancer) that are sufficiently specific, sensitive, fast, and economically sustainable.

An ideal screening platform should be rapid, sensitive, specific, robust, simple-to-use, and have sufficient throughput to be widely applicable in medical diagnostics. In addition, the determination of an analyte should preferably be carried out directly in tested samples (e.g., blood, plasma, urine, saliva, cerebrospinal fluid) with limited or no sample preparation. Diagnostic instruments allowing continuous monitoring of analyte concentration, which is not possible using conventional homogeneous and heterogeneous immunoassays, are also desirable. Biosensors present a promising alternative to established diagnostic technologies and can potentially meet many of these requirements. Surface plasmon resonance (SPR) biosensors offer a label-free direct measurement platform for rapid screening of medically relevant analytes. Recent advances in SPR sensor hardware, biorecognition elements and their immobilization, and sensor data analysis have made SPR biosensors a strong candidate for development of new analytical systems for medical diagnostics.

In the following section, we review the state of the art in applications of SPR biosensor technology for detection of disease biomarkers such as antigens and antibodies related to cancer, heart attack, and other diseases. Review of SPR applications in the field of hormone detection and monitoring of drug serum levels is also reported.

2
Cancer Markers

Early diagnosis is the key to successful treatment for most types of cancer. Conventional diagnostic methods such as X-ray imaging, computer tomography, or ultrasound are not appropriate for early stage cancer diagnostics because they detect already formed tumors. Detection of biological markers of cancer that are produced as the cancer grows is a helpful tool in cancer diagnostics and monitoring. For example, the prostate-specific antigen (PSA) – a biomarker of prostate cancer – can be detected in blood even before the cancer can be diagnosed by the conventional methods [5]. Diagnostic tests based on detection of biomarkers are non-invasive and less expensive than conventional methods such as biopsy of tissues (liver, kidneys, and testicles) and examinations involving mammography, ultrasonography, X-ray imaging instruments, or radiological and cytological devices. This makes cancer biomarker tests more applicable to large scale population screening, or repetitive screening of an individual [6].

There is a vast effort in research laboratories worldwide to identify new cancer biomarkers present in the circulatory system [7] and gene markers present in the human genome [8]. In the past two decades, hundreds of thousands of substances have been investigated as potential biomarkers for cancer diagnosis, but only about 50 serological tumor markers are currently available [9] and only 12 cancer biomarkers are recognized by the US Food and Drug Administration (FDA) [10]. Biomarkers of malignancy already in clinical use include PSA as the marker of prostate cancer [11, 12], carcinoembryonic antigen (CEA) marker (colorectal, breast cancer) [13, 14], cancer antigen (CA) marker CA 15-3 (breast cancer) [15], CA 125 (ovarian cancer) [16] carbohydrate antigen CA 19-9 (pancreas, colon, stomach cancer) [17], and alpha fetoprotein (AFP) marker (liver, testicular cancer) [18]. Many other potential cancer markers such as beta hCG (testicular cancer), calcitonin marker (thyroid cancer), and thyroglobulin marker (thyroid cancer) are under evaluation [8, 19, 20].

PSA is one of the most widely used cancer biomarkers. It is a chymotrypsin-like serine protease that is produced by epithelial cells of the prostate gland and secreted into the prostatic fluid. Prostate-cancer invasion disrupts the epithelial membrane barrier leading to elevated serum levels of PSA. Detection of PSA in blood can therefore be useful in the diagnosis of prostate abnormalities and for evaluation of prostate cancer therapy efficacy [21]. Two different forms of PSA are immunologically detectable: the free form (MW 34 kDa) and a complex with α-1-antichymotrypsin (MW 96 kDa). Diagnostic assays developed for detection of PSA (e.g., enzyme-linked immunosorbent assays) detect total PSA concentrations down to 0.1 ng mL^{-1} [22, 23].

An immunoassay for the detection of PSA in PBS buffer based on a dual-channel SPR instrument with angular modulation (IBIS II) has been reported [24]. This work compared direct and sandwich detection of PSA on planar- and hydrogel-type sensor surfaces. Amplification with colloidal gold and latex microspheres, respectively, was employed in the sandwich assay. Sensor chips with carboxylated matrices of different thicknesses were used. Mouse monoclonal antibodies against PSA were immobilized on the both types of chip surfaces via amine coupling chemistry. The first amplification step consisted of incubation with rabbit anti-PSA polyclonal antibodies. In experiments employing latex amplification, incubation with biotinylated goat anti-rabbit IgG was followed with streptavidin-coated latex microspheres. In experiments with gold microspheres, detection continued with immersion in a solution containing goat anti-rabbit IgG-coated colloidal gold. The chip with a thinner dextran matrix was found to lead to a higher sensor sensitivity for both the direct and sandwich detection formats. Specifically, detection limits of 0.15 ng mL^{-1} and 2.4 ng mL^{-1} were determined for the detections

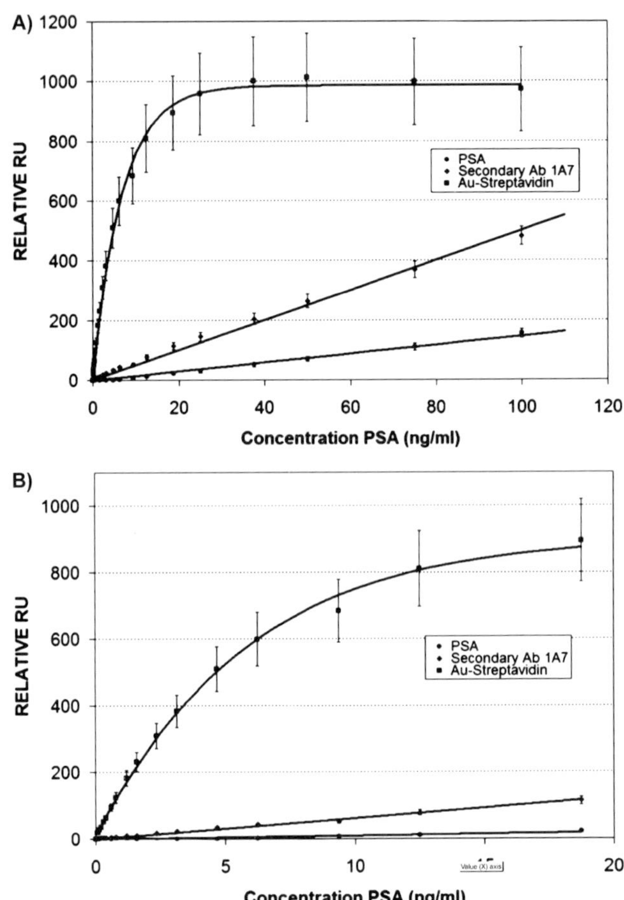

Fig. 1 Detection of PSA in buffer directly and using sandwich assay format. The signals generated upon binding of the different partners [PSA (● *bottom*), biotinylated secondary antibody (▲ *middle*) and 20 nm gold nanoparticles modified with streptavidin (■ *top*)] are shown for: **a** PSA concentrations varying between 73 pg mL^{-1} and 100 ng mL^{-1}; **b** PSA concentrations between 73 pg mL^{-1} and 20 ng mL^{-1} [25]

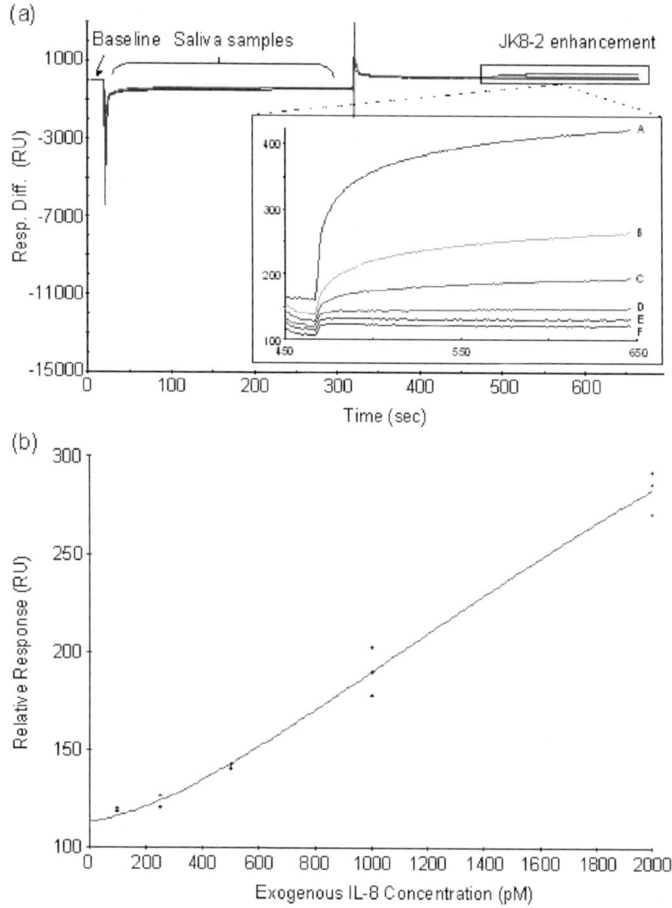

Fig. 2 Sensorgrams and calibration curve for different IL-8 concentrations in saliva supernatant premixed with $10\,\mathrm{mg\,mL^{-1}}$ of CM dextran sodium salt. **a** Sensorgrams corresponding to enhanced detection by secondary antibody for different IL-8 concentrations (5 nM, 2 nM, 1 nM, 500 pM, 250 pM, and 0 pM labeled as *A*, *B*, *C*, *D*, *E*, and *F*, respectively). **b** Calibration curve covering the IL-8 concentration range from 0 to 2 nM [26]

employing gold particle enhancement on planar-type and gel-type sensor surfaces, respectively.

Huang et al. investigated detection of PSA using direct and sandwich immunoassay formats using an SPR sensor Biacore 2000 [25]. PSA-receptor molecules consisting of a single domain antigen-binding fragment were covalently immobilized on the sensor surface via a mixed alkanethiol self-assembled monolayer (SAM). PSA concentrations as low as $10\,\mathrm{ng\,mL^{-1}}$ were detected in buffer. It was demonstrated that a sandwich assay involving a biotinylated secondary antibody and streptavidin-modified gold nanoparticles lowered the limit of detection for PSA below 1 ng mL^{-1}. Signal levels corresponding to direct and amplified PSA detections in buffer are displayed in Fig. 1.

Yang et al. [26] measured levels of interleukin-8 (IL-8) protein in the saliva of healthy individuals and patients with oropharyngeal squamous cell carcinoma using a Biacore X instrument. A sandwich assay using two monoclonal antibodies, which recognize different epitopes on the IL-8, was used. Monoclonal antibody against IL-8 was immobilized onto a dextran surface via amine coupling chemistry. Saliva samples were first centrifuged to clarify the supernatants. The supernatants were then aspirated and separated from the cellular pellet. The detection limit for IL-8 was determined to be 2.5 pM ($\sim$ 0.02 ng mL^{-1}) for detection in buffer and 184 pM ($\sim$ 1.5 ng mL^{-1}) for detection in saliva samples. Sensorgrams corresponding to IL-8 binding at different concentrations and a calibration curve for secondary antibody-enhanced detection of IL-8 (sandwich assay) are displayed in Fig. 2.

Nayeri et al. [27] presented an SPR (Biacore 2000)-based direct qualitative detection of hepatocyte growth factor (HGF), which is an angiogenic growth factor related to breast cancer, in reconstituted fecal samples from patients with infectious gastroenteritis ($n = 20$) and normal controls ($n = 10$) (dissolved in distilled water at a dilution rate of 1 : 6). Mouse anti-human HGF monoclonal antibodies and recombinant human HGF receptor were immobilized on a dextran surface. The proportion of antibody-positive patient samples detected by SPR correlated well with results obtained using ELISA.

3 Heart Attack Markers

Diagnosis of cardiac muscle injury relies on the detection of biomarkers such as troponin I (TnI), troponin C (TnC), myoglobin, fatty acid binding protein (FABP), glycogen phosporylase isoenzyeme BB (GPBB), C-reactive protein (CP), urinary albumin, creatine kinase myocardial band (CK-MB), and brain (B-type) natriuretic peptide in blood and urine [28–30].

Troponin complex is a heteromeric protein which plays an important role in the regulation of skeletal and cardiac muscle contraction. It consists of three subunits, troponin I (TnI), troponin T (TnT), and troponin C (TnC). Each subunit is responsible for part of troponin complex function. For more than a decade, the cardiac form of Tn I (cTn I) has been known as a reliable marker of cardiac tissue injury. The greatest advantage of detection of cTn I is its cardio-specificity [31].

Detection of human cTn I (29 kDa) in serum, utilizing direct and sandwich immunoassay formats, was carried out by an SPR sensor with wavelength modulation [32]. Biotinylated antibodies against cTn I were immobilized on an avidin layer created using amine coupling chemistry on an activated SAM. Sensor responses to cTn I binding to immobilized antibody in serum samples were compared with standard solutions containing known concentrations of cTn I. Two detection modes for cTn I were demonstrated: (1) direct detection of cTn I with a detection range of 2.5–40 ng mL^{-1}, and (2) a sandwich assay with a detection limit of 0.25 ng mL^{-1} and detection range of 0.5–20 ng mL^{-1}.

Detection of myoglobin and cTn I markers was carried out using a home-made two-channel multimode SPR fiber-optic sensor [33]. The two respective biomolecular recognition elements, human anti-myoglobin and human anti-cardiac troponin I, were immobilized on a dextran surfaces via amine coupling chemistry. Both myoglobin and cTnI were detected in buffer at concentrations lower than 3 ng mL^{-1}.

4 Antibodies

Antibodies are soluble proteins (immunoglobulins) secreted by B-lymphocytes in response to exogenous and endogenous antigens. Antibodies specifically bind to antigens to form antigen–antibody complexes. Antigens in this complex are typically inactive and thus interaction of antigen with other host molecules is blocked. These antibodies are referred to as neutralizing or inhibiting antibodies. In contrast, there are antibodies with a stimulating effect, i.e., they activate bound molecules. An example of disease caused by activating antibody is Grave's disease when antibodies function as ligands to cell receptors. Presence of specific antibodies in the circulatory system can thus serve as a biomarker of various diseases such as microbial infection, virus infection, allergy, autoimmune disease, or tissue injury.

Direct detection of antibodies against Epstein–Barr virus (anti-EBNA) in 1% human serum was carried out using a wavelength-modulated SPR biosensor (sensor setup description in [34, 35]). Synthetic peptides were used as receptors and immobilized on the sensor surface in the form of BSA–peptide conjugates via hydrophobic and electrostatic interactions [36]. A sensor calibration curve was established for an anti-EBNA concentration range of 0.2–2000 ng mL^{-1} (Fig. 3). The sensor response showed reproducibility better than 82% for all concentrations and multiple chips and over 90% for measurements performed on a single chip. The lowest detection limit for the direct detection of anti-EBNA was found to be 0.2 ng mL^{-1}. A procedure for regeneration of the sensor was developed and was demonstrated to allow at least 10 repeated anti-EBNA detection experiments without a significant loss in sensor sensitivity. In addition, it was demonstrated that the sensor chips can be stored for 30 days without deterioration in performance (Fig. 4) [37].

The presence of antibodies against human respiratory syncytial virus (RSV) in 26 patient sera was detected using an SPR biosensor (Biacore 2000)

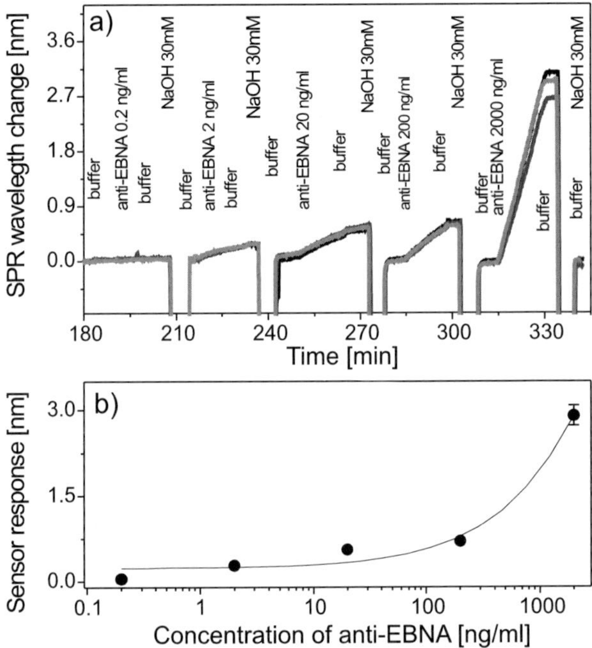

Fig. 3 **a** Sensor response to anti-EBNA detection obtained from three individual sensing channels on regenerated surface. **b** Sensor calibration curve [37]

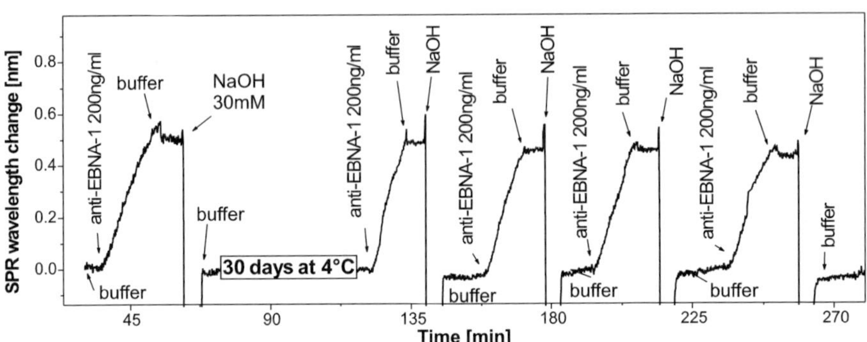

Fig. 4 Sensor responses to anti-EBNA binding obtained using one sensor chip immediately after functionalization and after 30 days of storage [37]

by McGill et al. [38]. Monoclonal antibodies against the virus glycoproteins (F- and G-glycoproteins) were covalently attached to a dextran matrix via amine coupling chemistry and then used to immobilize the respective virus glycoproteins. Serum samples isolated from patient respiratory tracts were diluted in HBS buffer (1 : 10), filtered (0.22 μm filter), and injected into the sensor surfaces. It was shown that, in contrast to an immunofluorescence assay, the SPR biosensor was capable of recognizing the antigenic differences between the two different contemporary genotypes of the virus (G- and F- virus glycoproteins). In order to confirm that the material binding to virus antigen was immunoglobulin, monoclonal mouse anti-human IgG was injected after each serum sample. In all SPR measurements the detection of antibodies was genotype specific.

Isotype-specific anti-adenoviral antibodies in patients dosed with adenoviral-based gene therapy vectors were detected using a Biacore 3000 by Abad et al. [39]. In this assay, whole intact virus was immobilized on the sensor surface (dextran matrix) using amine coupling chemistry. The binding of antibodies from patient blood sera (1 : 10 diluted) was measured by SPR sensor and ELISA. The results obtained by the SPR biosensor were consistent with those obtained using ELISA.

Direct qualitative detection of antibodies against hepatitis G virus from patient sera was presented by group the of Rojo [40]. A Biacore 1000 immunoassay utilized synthetic peptides, which were immobilized on dextran surface via amine coupling chemistry. Sera from 38 chronic hepatitis C patients, 36 hemodialyzed patients and 110 control healthy individuals (1 : 100 dilution) were tested for the presence of specific antibodies. The results were in good agreement with those obtained using ELISA.

The detection of antibodies against herpes simplex virus type 1 and type 2 (HSV-1, HSV-2) in 1 : 100 diluted human sera with HBS buffer using the Biacore X instrument is reported in the work of Wittekindt et al. [41]. Peptides, used as a biorecognition element, were biotinylated and immobilized on the streptavidin-coated sensor chip. Two peptides from a series of eight peptides selected from segments of HSV-1 and HSV-2 gB, respectively were identified as immunogenic. Employing both peptides, a good agreement between the SPR biosensor and immunoblotting (reference method) was obtained (correlation 83% and 86% for antibodies against HSV-1 and HSV-2, respectively).

An angular modulation-based SPR biosensor for syphilis screening has been reported by Severs et al. [42]. Antibodies against the causative organism *Treponema pallidum* were detected in serum (1 : 20 diluted in Tris buffer) using sensor chips coated with recombinant *Treponema pallidum* membrane protein A (TmpA) and blocked with 0.1% gelatin. Direct and sandwich assay formats were used for detection. It was shown that the direct detection of antibodies in serum was not sufficiently reproducible, most likely due to non-uniformity of patient serum samples. In contrast, the results obtained with the SPR sandwich assay for ten blind-coded sera corresponded well with conventional syphilis tests (*Trepanema pallidum* haemagglutination assay, fluorescent treponemal antibody-absorbed test, venereal diseases research laboratory flocculation test, and TmpA-based ELISA test).

Monitoring plasma levels of anti-protein S antibodies following Varicella–Zolter virus infection has been reported by Regnault et al. [43], who used an SPR sensor (Biacore X) to detect the presence of antibodies to protein S in infected patients [44]. Immobilization of protein S and IgG, respectively on dextran surfaces was carried out using amine coupling chemistry. Direct qualitative detection of antibodies was performed in diluted plasma (1 : 5) of patient sera samples each day during a 45-day infection. Twelve plasma samples from healthy patients were used as a control for potential non-specific binding.

Direct and sandwich format detection of antibodies against glucose 6-phosphate isomerase (GPI) in synovial fluids of rheumatoid arthritis and osteoarthritis patients (diluted 1 : 100 in HEPES) using a Biacore 2000 is presented in the work of Kim et al. [45]. Recombinant human GPI proteins produced from *E. coli* were immobilized on the dextran sensor surface via amine coupling chemistry. The synovial fluid samples from rheumatoid arthritis patients showed a significantly higher level of antibodies binding to the recombinant GPI proteins than samples from osteoarthritis patients.

Wilkop et al. reported the use of an SPR imaging sensor and a micro-contact printed array for parallelized detection of antibodies against cholera toxin (CT) [46]. Immobilization of the toxin was performed by combining the microprinting method with the covalent linkage of the protein to NHS-activated terminal groups on a self-assembled monolayer of thiols. Detection of anti-cholera toxin IgG (anti-CT IgG) was demonstrated for antibody concentrations ranging from 10 to 100 $\mu g\,mL^{-1}$, Fig. 5.

An assay for diagnosing type I diabetes mellitus based on the detection of anti-glutamic acid decarboxylase (GAD) antibodies in buffer by a Biacore 2000, is presented in [47, 48]. Biotinylated GAD was immobilized on a streptavidin-coated surface. The effect of mixed SAM composition (differing in ratios of hydroxyl- and carboxyl-terminated alkanethiols) on the sensitivity of the sensor was investigated. On SPR sensor chips prepared with the optimized SAM composition (10 : 1 ratio of 3-mercaptopropanol to 11-mercaptoundecanoic acid), a concentration of anti-GAD as low as $0.75\,\mu g\,mL^{-1}$ was detected.

An SPR sensor-based detection of antibodies against granulocyte macrophage colony stimulating factor (GM-CSF) was performed in the work of Rini et al. [1]. GM-CSF is cytokine that is involved in human immunotherapy protocols for various cancers including prostate cancer [49]. Antibodies against GM-CSF were induced in prostate cancer patients by repeated administration of GM-CSF and their presence in patient sera was monitored using a Biacore 2000 (sera diluted 1 : 5) and ELISA (sera diluted 1 : 20). The GM-CSF antigen used as a biomolecular recognition element was immobilized on the carboxymethylated dextran on the surface of the SPR sensor via amine coupling chemistry. The measurements performed using the SPR biosensor revealed that all 15 prostate cancer patients treated with GM-CSF produced GM-CSF reactive antibodies, which was in agreement with reference ELISA measurements.

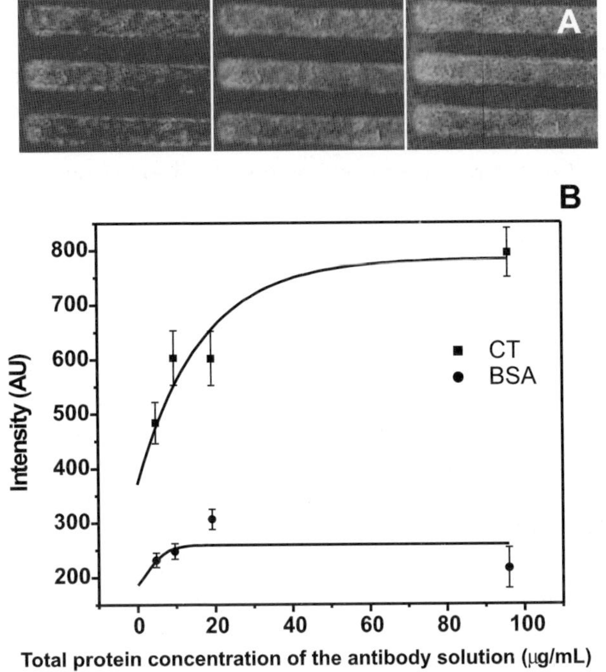

Fig. 5 Detection of anti-CT antibodies with printed CT proteins. **a** Images of printed CT patterns before and after incubation with increasing amount of anti-CT antibodies (background corrected). From *left* to *right*: sensor surface before incubation with anti-CT; after exposure to anti-CT (0.001 $mg\,mL^{-1}$); and after exposure to anti-CT (0.002 $mg\,mL^{-1}$) **b** Concentration-dependent response to anti-CT binding to CT (measuring surface) and BSA (reference surface), respectively [46]

Direct detection of antibody against insulin in patient sera using an SPR sensor Biacore 2000 is presented in [50]. Purified human insulin was used as a biorecognition element and immobilized on the sensor surface via amine coupling chemistry. Test sera samples were pretreated to remove insulin and filtered before SPR measurements. Insulin antibodies were detected in eight selected patient sera samples and fell in the range 2.91–16.3 $\mu g\,mL^{-1}$.

5 Hormones

Monitoring concentrations of female hormones is important for female disease diagnostics as well as for fetal health monitoring. The most important female cycle biomarkers, which are typically measured in clinical laboratory tests or commercial test strips, include follicle stimulating hormone (FSH) as a marker of non-pregnancy, luteinizing hormone (LH) as a marker of ovulation, and human chorionic gonadotropin hormone (hCG) as a marker of pregnancy.

Direct detection of hCG in buffer in the concentration range 0.05–1 μg mL^{-1} was demonstrated using a wavelength modulation-based SPR sensor, Fig. 6. A regeneration protocol was developed that allowed repeated use of the sensor with measurement reproducibility over 90%. The same group used an SPR imaging instrument with antibodies against hCG covalently immobilized to a mixed SAM to detect hCG at concentrations lower than 500 ng mL^{-1} [51].

Ladd et al. reported SPR sensor-based detection of hCG [52], exploiting a DNA-directed antibody immobilization method. The immobilization consisted of non-covalent attachment of streptavidin to a biotinylated SAM followed by binding of biotinylated oligonucleotides to available streptavidin binding sites. Antibodies chemically modified with oligonucleotides with a complementary sequence were finally attached to this surface via DNA hybridization. The detection limit for direct detection of hCG in buffer by a dual-channel SPR sensor with wavelength modulation was determined to be 0.5 ng mL^{-1}.

Detection of estrone and estradiol in buffer using SPR sensors Biacore 2000 and Biacore 1000 was carried out by Coille et al. [53]. Analyte–BSA conjugates and BSA were immobilized in the sensing and reference channels of a sensor chip, respectively, using NHS-esters. Analyte concentrations detected using inhibition format in this work were in the range 0.01–3000 ng mL^{-1}.

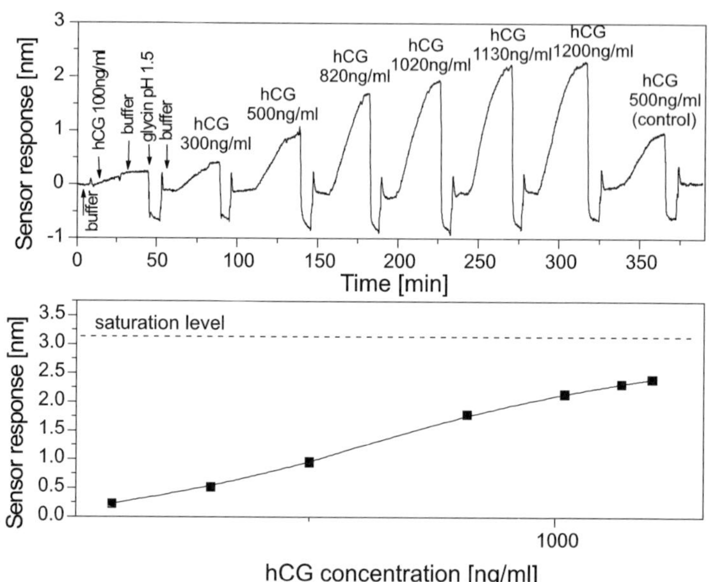

Fig. 6 Sensorgrams corresponding to binding of hCG at various concentration on anti-hCG immobilized on the sensor surface and hCG-sensor calibration curve

Miyashita et al. [54] present an SPR Biacore X-based immunoassay for the detection of 17β-estradiol in buffer. The assay was performed in an inhibition format, in which 17β-estradiol BSA conjugates were immobilized on the sensor surface and the binding of antibody to 17β-estradiol conjugates on the surface was measured. The 17β-estradiol was detected in the concentration range 0.47–21.4 nM ($\sim$ 0.14–6.4 ng mL^{-1}).

6 Drugs

Therapeutic drug monitoring is very important for treatment of many serious diseases (e.g., HIV [55], heart failure [56], Parkinson's, malaria, cystic fibrosis, diabetes mellitus [57], etc.) and for treatment of pregnant women, children, and patients with special conditions (e.g., pre-existing liver damage) where routine recommended dosing is not always appropriate.

An inhibition assay for the detection of anticoagulatory coumarin derivative 7-hydroxycoumarin (7-OHC) using a Biacore SPR sensor and competitive assay was presented by Keating et al. [58]. A 7-OHC conjugated with BSA was immobilized on a carboxymethylated dextran sensor chip via amine coupling chemistry. Serum samples (diluted with buffer) were premixed with a polyclonal anti-7-OHC antibody and injected over the sensor surface. The binding of excess antibodies to the immobilized conjugate generated a sensor response inversely proportional to the 7-OHC concentration. The assay had a measuring range of 0.5–80 µg mL^{-1}. This immunoassay exhibited reproducibility and sensitivity comparable to established methods of analysis.

Fitzpatrick et al. detected oral anticoagulant warfarin using a Biacore 3000 SPR sensor and an inhibition assay [59]. 4′-Aminowarfarin or 4′-azowarfarin–BSA was immobilized on a dextran matrix via amine coupling chemistry. Detection of warfarin was performed in plasma ultrafiltrate (diluted 1 : 100) in a concentration range of 4–250 ng mL^{-1}. The observed calibration curve and residual plot are shown in Fig. 7. A procedure for regeneration of the sensor chips was established, allowing for more than 70 binding cycles.

Direct detection of a cytokine protein, recombinant human interferon-γ, using an IBIS SPR sensor is presented in [60]. Several types of sensor chip coatings, including self-assembled monolayers and hydrogel-derivatized SAMs, were characterized in terms of their ability to resist non-specific adsorption from plasma. The best results with respect to plasma adsorption and surface regenerability were obtained with antibodies immobilized on the dextran-modified 11-mercaptoundecanoic acid SAM. The detection limit for detection of human interferon-γ in 1 : 100 diluted plasma was established at 250 ng mL^{-1}.

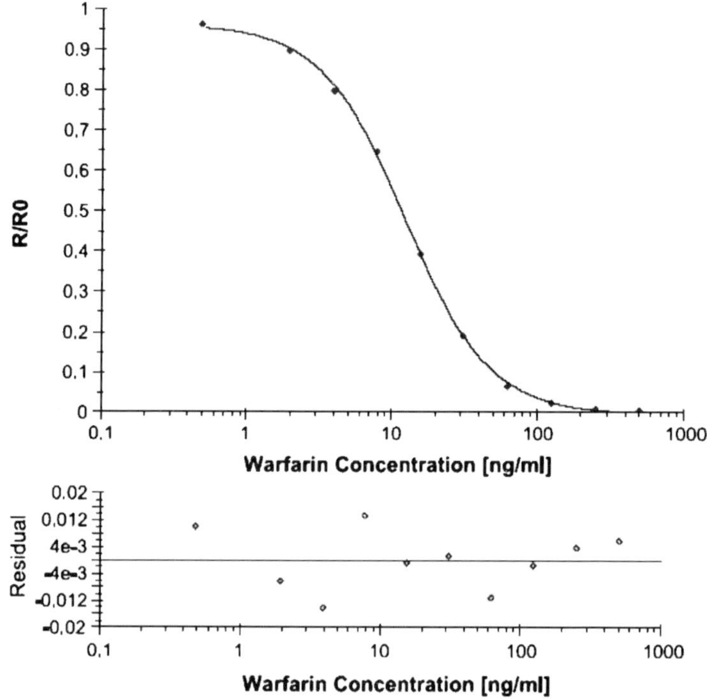

Fig. 7 Calibration and residual plot for the inhibition assay-based detection of warfarin in plasma ultrafiltrate ($n = 3$). The mean normalized response value (RAG/R0) at each analyte concentration from three independent assays was used to calculate the calibration curve and to determine the assay variation [59]

Detection of low molecular weight heparin oligosaccharide (Fragmin), which is an antithrombotic agent, was demonstrated using purified monoclonal antibodies immobilized via amine coupling chemistry onto the surface of a Biacore 3000 [61]. Monoclonal antibodies were immobilized on a sensor surface prefunctionalized with Fragmin–HSA conjugates. The detection limit for Fragmin in PBS buffer was determined to be 125 nM ($\sim$ 625 ng mL^{-1}).

The Sakai group [62] developed an SPR sensor-based inhibition immunoassay for detection of morphine in buffer and in 1% human urine. They used an SPR sensor SPR-20 and immobilized morphine–BSA conjugates on the gold surface via physical adsorption. The addition of morphine to the anti-morphine antibody solution was found to reduce the SPR signal because of the inhibition effect of morphine. The detection limit of morphine in 1% urine was established at 2 ng mL^{-1}.

A Biacore 1000 inhibition assay for the detection of morphine-3-glucuronide (M3G), the main metabolite of heroin and morphine, was demonstrated in buffer and diluted urine by Dillon et al. [63]. M3G–ovalbumin conjugate was

Table 1 Overview of biomarkers related to medical diagnostics detected with SPR biosensors

Biomarker	Disease	Normal level	Detection limit [ng mL^{-1}]	BRE, detection format	Sample	Refs.
Cancer biomarkers:						
PSA	Prostate cancer	0–4 ng mL^{-1}	0.15	Antibody, enhanced detection by gold nanoparticles conjugated with antibody	Buffer containing 3% BSA	[24]
PSA	Prostate cancer	0–4 ng mL^{-1}	10	PSA-specific receptor, direct	Buffer	[25]
PSA	Prostate cancer	0–4 ng mL^{-1}	< 1	PSA-specific receptor, sandwich detection enhanced with gold nanoparticles	Buffer	[25]
Interleukin-8	Oropharyngeal cancer		0.02	Antibody, sandwich	Buffer	[26]
Interleukin-8	Oropharyngeal cancer		1.5	Antibody, sandwich	Saliva (diluted)	[26]
Heart attack markers:						
Troponin I	Cardiac tissue injury	0–0.1 ng mL^{-1}	2.5	Antibody, direct	Serum	[32]
Troponin I	Cardiac tissue injury	0–0.1 ng mL^{-1}	0.25	Antibody, sandwich	Serum	[32]
Troponin I	Cardiac tissue injury	0–0.1 ng mL^{-1}	< 3	Antibody, direct	Buffer	[33]
Myoglobin	Cardiac tissue injury	0–85 ng mL^{-1}	< 3	Antibody, direct	Buffer	[33]
Antibodies:						
Antibody against Epstein–Barr virus	Mononucleosis		0.2	BSA–peptide conjugate, direct	Serum (1%)	
Antibody against cholera toxin			10 000	Antigen, direct (SPR imaging)	Buffer	[46]
Antibody against GM-CSF	Prostate cancer			GM-CSF antigen, direct	Serum (diluted 1 : 5 with buffer)	[1]
Antibody against insulin	Diabetes mellitus		2910	Antigen, direct	Serum (pretreated to remove insulin)	[50]

Table 1 (continued)

Biomarker	Disease	Normal level	Detection limit [ng mL^{-1}]	BRE, detection format	Sample	Refs.
Hormones:						
hCG	Female hormonal diseases/ pregnancy marker	Normal: 0–10 U/L Pregnancy: > 500 U/L	50	Antibody, direct	Buffer	
hCG	Female hormonal diseases/ pregnancy marker	Normal: 0–10 U/L Pregnancy: > 500 U/L	500	Antibody, direct	Buffer	[51]
hCG	Female hormonal diseases/ pregnancy marker	Normal: 0–10 U/L Pregnancy: > 500 U/L	0.5	Antibody, sandwich	Buffer	[52]
17β-Estradiol	Hormonal diseases	M: 0.02–0.05 F (0–55 years): 0.01–0.5 F (55–110 years): 0–0.04	0.14	Antibody, inhibition	buffer	[54]
Drugs:						
Warfarin			4	Antibody, inhibition	Plasma ultrafiltrate (diluted 1 : 100)	[59]
interferon-γ			250	Antibody, direct	Spiked plasma (1 : 100)	[60]
Fragmin			625	Antibody, direct	Buffer	[61]
Morphine-3-glucuronide			0.7	Antibody, inhibition	Urine (1 : 250 diluted)	[63]
Morphine			2	Antibody, inhibition	Urine (1%)	[62]

SPR Surface plasmon resonance, *BRE* biorecognition elements, *PSA* prostate-specific antigen, *BSA* bovine serum albumin, *GM-CSF* granulocyte macrophage colony stimulating factor, *hCG* human chorionic gonadotropin immobilized on a dextran matrix via amine coupling chemistry. Two polyclonal antibodies were produced, purified, and tested for M3G detection. The range of detection of M3G in buffer and in urine (diluted 1 : 250) was found to be 0.7 and 24.4 ng mL^{-1}, respectively.

7 Summary

Recently, we have witnessed an increasing effort to exploit SPR biosensor technology for medical diagnostics. Detection of a variety of disease biomarkers, hormones, and drugs at clinically relevant levels has been demonstrated. Although many of these detection experiments were performed in pure model samples with minimal or no matrix interferences, clinical samples have also been tackled, Table 1.

It is expected that advances in SPR sensor instrumentation (reducing size, improving sensitivity, increasing throughput), biorecognition elements and methods for their immobilization (increasing sensitivity and specificity) will lead to new systems for rapid detection and identification of disease biomarkers. These will further extend the applicability of SPR biosensor technology in medical diagnostics.

References

1. Rini B, Wadhwa M, Bird C, Small E, Gaines-Das R, Thorpe R (2005) Cytokine 29:56
2. Luppa PB, Sokoll LJ, Chan DW (2001) Clin Chim Acta 314:1
3. McGlennen RC (2001) Clin Chem 47:393
4. Peter C, Meusel M, Grawe F, Katerkamp A, Cammann K, Borchers T (2001) Fresenius J Anal Chem 371:120
5. Landman J, Chang Y, Kavaler E, Droller MJ, Liu BC (1998) Urology 52:398
6. Hlawatsch A, Teifke A, Schmidt M, Thelen M (2002) Am J Roentgenol 179:1493
7. Canto EI, Shariat SF, Slawin KM (2004) Curr Urol Rep 5:203
8. Yasui W, Oue N, Ito R, Kuraoka K, Nakayama H (2004) Cancer Sci 95:385
9. Wu J (2001) Diagnosis and management of cancer using serologic tumor markers. Saunders, Philadelphia
10. Anderson NL, Anderson NG (2002) Mol Cell Proteomics 1:845
11. Schmid HP, Riesen W, Prikler L (2004) Crit Rev Oncol Hematol 50:71
12. Caplan A, Kratz A (2002) Am J Clin Pathol 117:S104
13. Goldstein MJ, Mitchell EP (2005) Cancer Invest 23:338
14. Duffy MJ (2001) Clin Chem 47:624
15. Duffy MJ, Duggan C, Keane R, Hill AD, McDermott E, Crown J, O'Higgins N (2004) Clin Chem 50:559
16. Fehm T, Heller F, Kramer S, Jager W, Gebauer G (2005) Anticancer Res 25:1551
17. Micke O, Bruns F, Schafer U, Kurowski R, Horst E, Willich N (2003) Anticancer Res 23:835

18. Daniele B, Bencivenga A, Megna AS, Tinessa V (2004) Gastroenterology 127:S108
19. Diamandis EP (2004) Clin Chem 50:793
20. Marrero JA, Lok AS (2004) Gastroenterology 127:S113
21. Polascik TJ, Oesterling JE, Partin AW (1999) J Urol 162:293
22. Kuriyama M, Wang MC, Papsidero LD, Killian CS, Shimano T, Valenzuela L, Nishiura T, Murphy GP, Chu TM (1980) Cancer Res 40:4568
23. Acevedo B, Perera Y, Ruiz M, Rojas G, Benitez J, Ayala M, Gavilondo J (2002) Clin Chim Acta 317:55
24. Besselink GA, Kooyman RP, van Os PJ, Engbers GH, Schasfoort RB (2004) Anal Biochem 333:165
25. Huang L, Reekmans G, Saerens D, Friedt JM, Frederix F, Francis L, Muyldermans S, Campitelli A, Hoof CV (2005) Biosens Bioelectron 21:483
26. Yang CY, Brooks E, Li Y, Denny P, Ho CM, Qi FX, Shi WY, Wolinsky L, Wu B, Wong DTW, Montemagno CD (2005) Lab on a Chip 5:1017
27. Nayeri F, Aili D, Nayeri T, Xu JY, Almer S, Lundstrom I, Akerlind B, Liedberg B (2005) BMC Gastroenterol 5:13
28. McDonough JL, Van Eyk JE (2004) Prog Cardiovasc Dis 47:207
29. Thielmann M, Massoudy P, Marggraf G, Knipp S, Schmermund A, Piotrowski J, Erbel R, Jakob H (2004) Eur J Cardiothorac Surg 26:102
30. Matveeva EG, Gryczynski Z, Lakowicz JR (2005) J Immunol Methods 302:26
31. Bodor GS, Porterfield D, Voss EM, Smith S, Apple FS (1995) Clin Chem 41:1710
32. Wei J, Mu Y, Song D, Fang X, Liu X, Bu L, Zhang H, Zhang G, Ding J, Wang W, Jin Q, Luo G (2003) Anal Biochem 321:209
33. Masson JF, Obando L, Beaudoin S, Booksh K (2004) Talanta 62:865
34. Dostálek J, Vaisocherová H, Homola J (2005) Sensor Actuator B Chem 108:758
35. Homola J, Vaisocherová H, Dostálek J, Piliarik M (2005) Methods 37:26
36. Houska M, Brynda E, Bohatá K (2004) J Colloid Interface Sci 273:140
37. Vaisocherová H, Mrkvová K, Piliarik M, Jinoch P, Šteinbachová M, Homola J (2006) Biosens Bioelectron, in press
38. McGill A, Greensill J, Marsh R, Craft AW, Toms GL (2004) J Med Virol 74:492
39. Abad LW, Neumann M, Tobias L, Obenauer-Kutner L, Jacobs S, Cullen C (2002) Anal Biochem 310:107
40. Rojo N, Ercilla G, Haro I (2003) Curr Protein Pept Sci 4:291
41. Wittekindt C, Fleckenstein B, Wiesmuller K, Eing BR, Kuhn JE (2000) J Virol Methods 87:133
42. Severs AH, Schasfoort RBM, Salden MHL (1993) Biosens Bioelectron 8:185
43. Regnault V, Boehlen F, Ozsahin H, Wahl D, de Groot PG, Lecompte T, de Moerloose P (2005) J Thromb Haemost 3:1243
44. Levin M, Eley BS, Louis J, Cohen H, Young L, Heyderman RS (1995) J Pediatr 127:355
45. Kim JY, Lee MH, Jung KI, Na HY, Cha HS, Ko EM, Kim TJ (2003) Exp Mol Med 35:310
46. Wilkop T, Wang Z, Cheng Q (2004) Langmuir 20:11141
47. Choi SH, Lee JW, Sim SJ (2005) Biosens Bioelectron 21:378
48. Lee JW, Sim SJ, Cho SM, Lee J (2005) Biosens Bioelectron 20:1422
49. Mellstedt H, Fagerberg J, Frodin JE, Henriksson L, Hjelm-Skoog AL, Liljefors M, Ragnhammar P, Shetye J, Osterborg A (1999) Curr Opin Hematol 6:169
50. Kure M, Katsura Y, Kosano H, Noritake M, Watanabe T, Iwaki Y, Nishigori H, Matsuoka T (2005) Intern Med 44:100
51. Piliarik M, Vaisocherová H, Homola J (2005) Biosens Bioelectron 20:2104
52. Ladd J, Boozer C, Yu Q, Chen S, Homola J, Jiang S (2004) Langmuir 20:8090
53. Coille I, Gauglitz G, Hoebeke J (2002) Anal Bioanal Chem 372:293

54. Miyashita M, Shimada T, Miyagawa H, Akamatsu M (2005) Anal Bioanal Chem 381:667
55. Cooley LA, Lewin SR (2003) J Clin Virol 26:121
56. Dunselman PHJM, Scaf AHJ, Kuntze CEE, Lie KI, Wesseling H (1988) Eur J Clin Pharmacol 35:461
57. Lin EHB, Katon W, Von Korff M, Rutter C, Simon GE, Oliver M, Ciechanowski P, Ludman EJ, Bush T, Young B (2004) Diabetes Care 27:2154
58. Keating GJ, Quinn JG, O'Kennedy R (1999) Anal Lett 32:2163
59. Fitzpatrick B, O'Kennedy R (2004) J Immunol Methods 291:11
60. Stigter EC, Jong GJ, van Bennekom WP (2005) Biosens Bioelectron 21:474
61. Liljeblad M, Lundblad A, Ohlson S, Pahlsson P (1998) J Mol Recognit 11:191
62. Sakai G, Ogata K, Uda T, Miura N, Yamazoe N (1998) Sensor Actuator B Chem 49:5
63. Dillon PP, Daly SJ, Manning BM, O'Kennedy R (2003) Biosens Bioelectron 18:217

Subject Index

Printing: Krips bv, Meppel
Binding: Stürtz, Würzburg